Cognitive Radio and Dynamic Spectrum Access

Cognitive Radio and Dynamic Spectrum Access

Lars Berlemann
Deutsche Telekom, Germany

and

Stefan Mangold
Swisscom, Switzerland

A John Wiley and Sons, Ltd, Publication

This edition first published 2009
© 2009 John Wiley & Sons Ltd.

Registered office
John Wiley & Sons Ltd, The Atrium, Southern Gate, Chichester, West Sussex,
PO19 8SQ, United Kingdom

For details of our global editorial offices, for customer services and for information about how to apply for
permission to reuse the copyright material in this book please see our website at www.wiley.com.

Library of Congress Cataloging-in-Publication Data

Berlemann, Lars.
  Cognitive radio and dynamic spectrum access / Lars Berlemann, Stefan Mangold.
    p. cm.
  Includes bibliographical references and index.
  ISBN 978-0-470-51167-1 (cloth)
  1. Cognitive radio networks. 2. Radio frequency allocation. 3. Software radio.
  I. Mangold, Stefan. II. Title.
  TK5103.4815.B47 2008
  621.384—dc22

                                                          2008041820

A catalogue record for this book is available from the British Library.

ISBN 978-0-470-51167-1 (H/B)

Set in 10/12pt Times by Integra Software Services Pvt. Ltd, Pondicherry, India
Printed in Great Britain by CPI Antony Rowe, Chippenham, Wiltshire

Contents

List of Figures

List of Tables

About the Authors

Lars Berlemann and Stefan Mangold both contribute to research programs, standardization and industry innovations in the field of dynamic spectrum access, cognitive radio, and IEEE 802 standards. Together they have filed numerous journal and magazine articles, patents, and contributions to research conferences and workshops. They both consult government organizations such as the European Commission in steering related research programs. Lars Berlemann and Stefan Mangold have been delivering tutorials on cognitive radio at various research conferences such as IEEE PIMRC and the European Wireless Conference.

As alumni from RWTH Aachen University, Germany, Lars Berlemann and Stefan Mangold graduated at the Chair of Communication Networks, ComNets, with Professor Bernhard H. Walke as PhD advisor. Their PhD theses (both awarded *summa cum laude*) are today considered to be important early research contributions to the field and are in the highest ranks in the number of downloads from the University's download servers. Together with Professor Walke, Drs Berlemann and Mangold coedited the Wiley book *IEEE 802 Wireless Systems: Protocols, Multi-Hop Mesh/Relaying, Performance and Spectrum Coexistence* which was published in November 2006.

Lars Berlemann is product manager in the product and innovation department of Deutsche Telekom, Germany. Stefan Mangold is manager at Swisscom, Switzerland, leading the access team of the product IT development group of Swisscom Network & IT. They both work for companies that operate mobile, fixed, and broadcast networks, and in addition provide services with excellent customer focus. Consequently, Drs Berlemann and Mangold understand and exploit the synergies between academic research focusing on excellence, and industry innovations focusing on commercial exploitation.

Lars Berlemann and Stefan Mangold share and disseminate what they learn. In parallel with their employment, they enjoy working with students. Lars Berlemann is guest lecturer at the Chair of Communication Networks, Technical University of Dortmund, Germany. Stefan Mangold is with ETH Zurich, Switzerland, Department of Computer Science, where he works as lecturer and visiting scientist.

In addition to his scientific engineering background, Lars Berlemann holds a diploma in Business and Economics from RWTH Aachen.

The comments and statements made in this book are from the authors and do not necessarily reflect the official position of their employers.

Foreword

In the relatively crowded collection of books on cognitive radio, Stefan and Lars have made a unique and important contribution. First, there is no other single desk reference with such a comprehensive treatment of the applications of game theory to the many aspects of cognitive radios addressed here.

Along with a cutting-edge treatment of the rapidly evolving standards, their insightful treatment of horizontal and vertical behaviors of cognitive radios stands out. Horizontal etiquettes apply to the potential for cooperative sharing across bands, such as the ISM bands, where today, each user is both a primary and secondary user – first come first served. In licensed bands, cognitive behavior enhances secondary usage, and this can include handsets and cell sites, specifically authorized devices such as femtocells, and in-band *ad hoc* networks that fill gaps opportunistically. This book shows how cognitive behavior improves overall spectrum efficiency for both licensed and unlicensed spectrum management regimes, a significant contribution indeed.

The overall treatment goes beyond the anecdotal, developing the mathematical foundations with a rigor worthy of the best archival journals. In addition, Lars' in-depth treatment and extensions to the DARPA XG policy language build solidly on the XG foundation and, more importantly, go beyond the mere framework of XG to realize a substantially complete engineering basis for transitioning the policy language from US DoD's XG niche into a powerful tool for realizing practical commercial, military, and first-responder cognitive wireless networks.

So it seems clear that this book will contribute to the art and science of cognitive radio both immediately, in its comprehensive approach to emerging standards, and long term from its insightful treatment of the mathematical foundations.

Joe
Joseph Mitola III
Stevens Institute of Technology

Preface

When we started our joint research work in 2002 with the analysis of the competition for spectrum access of coexisting Wi-Fi networks in unlicensed spectrum we did not expect that soon a fast growing and highly successful society of interdisciplinary researchers would emerge. Researchers from electrical engineering, computer science, political economics and law represent today the base of the global cognitive radio research community. Now, after nearly a decade of various academic efforts, we are enjoying observing the well established research community with many funding programs, conferences and multiple special issues of magazines and journals. Initial steps towards standardization of cognitive radio and dynamic spectrum access have been reached. Many regulation authorities have already formed their opinion on cognitive radio – resulting in manifold positions and assessments of cognitive radio and its underlying economic potential.

Radio spectrum, and the right to exploit radio resources commercially are in the economic interest of wireless network operators. It is usually part of their long term business strategy: Changes in the way we regulate and license radio spectrum are closely monitored. It is no surprise that cognitive radio and dynamic spectrum access have been carefully examined by stakeholders of the Telecommunication, Information, Multimedia and Entertainment (TIME) industry from the very beginning. Today, the technological neutral spectrum licensing and a more flexible licensing approach enabling fast access to additional spectrum are seen as a benefit, especially when network coverage and capacity have to be increased. Even more advanced concepts such as the true cognitive radio may in the future enable a support of quality-of-service in shared spectrum, which is again praised by regulators and network operators. Under-utilized spectrum can be exploited with the concepts of dynamic spectrum access and cognitive radio. Access to new spectrum can be simplified, but also thresholds for new market entrants are essentially under a more closer control of radio regulators.

All such benefits justify a closer look at cognitive radio and its implications on the technology roadmaps of infrastructure manufacturers, the business strategies of network operators and the general impact on our global information society. After the start as an academic curiosity and the hype in the recent years we would like to contribute with this book to an initial assessment of advancements and maturity in research on cognitive radio and dynamic spectrum access. Especially with our introduction of operator-assisted cognitive radio we would like to highlight the emerging business opportunities from cognitive radio not only for new market entrants but also for incumbent wireless network operators.

We would be pleased if we could help to vitalize the discussion on cognitive radio in research and standardization, and if we could help establishing more trust in cognitive radio technology in spectrum regulation. But we also would like to highlight the opportunities and threats for incumbent operators arising from the inevitable introduction of spectrum liberalization and radio access technologies implementing dynamic spectrum access. Additionally, we condense some of our conference and journal publications related to Wi-Fi & WiMAX, mesh & multi-hop & relay-based networks, coexistence, Quality-of-Service support, spectrum sharing, spectrum etiquette & policies, flexible protocol stacks, dynamic spectrum access and cognitive radio of the last years to make it accessible to a general readership and mainstream research.

This text book is written for advanced undergraduate and graduate students as well as for practicing engineers and consultants in the telecommunication industry and government. Lecturers might take the content of this book as a basis and as accompanying literature for their courses on cognitive radio and future communication technologies. We assume that the reader of this book has a basic understanding of the principles of wireless communication. An interest in interdisciplinary research, an open mind for new solutions and a natural desire for questioning today's regime of spectrum regulation are also helpful when reading this book.

It is always difficult to satisfy the expectations of such a manifold readership. We therefore tried to balance between describing on the one hand initial steps towards practical realization and on the other hand the detailed introduction to our intensive scientific work towards the vision of true cognitive radio. It is a compromise between a general, tutorial-like introduction to cognitive radio and the scientific depth of the description of our meticulous research work.

With this textbook, we would like to give an outlook on the next key evolutions in wireless communications expected to be in parallel or beyond cognitive radios. In the close future, the principles of the Wi-Fi Orthogonal Frequency Division Multiplex (OFDM) physical layer will be adopted in future cellular radio networks based on LTE and maybe WiMAX. Basic OFDM will be enhanced to make it scalable and advanced antennas as well as MIMO will further increase the available capacity to the single user. The basic idea of dynamic spectrum access and cognitive radio principles might be also applied in LTE to increase availability of spectrum and the efficiency of spectrum usage. Another economically promising improvement of Wi-Fi will be 'Wi-Fi 2.0': a long-range Wi-Fi especially suited for fixed deployments in rural areas for provisioning wireless broadband access. 'Wi-Fi 2.0' is an exciting new approach to bridge coverage gaps in the provisioning of basic broadband services. It further is a cost-efficient alternative to expensive fixed networks based on copper or fiber.

In the midterm, 'Optical Wi-Fi' could become an attractive alternative to classical Wi-Fi systems for consumer electronics: Optical wireless communications is implemented for indoor communication at the wavelength of visible light taking the health concerns of today's users into account. Data transmissions are piggy-backed with the illumination lights for our living space. Such a consumer demand for healthy communication might lead to a revival of Infrared (IR) as radio interface for Wi-Fi. An increasingly important field for wireless communication will be the exploration of our universe: more and more satellites, probes, space crafts and stations (government owned as well as commercial) are sent out from earth having completely different requirements to wireless communication systems. 'Delay tolerant deep space communication' addresses such new application scenarios. For the time being, another related but more distant research field is the 'quantum communications,' which is a completely new discipline of research that could dwarf cognitive radio and promises revolutionary changes to wireless communication and information society.

We believe that it will still take decades of intensive research and development until our vision of a true cognitive radio will be realized and available to mass markets. Until then, our recommendation to decision makers of today's telecommunication industry on how to deal with cognitive radio depends on their participation in the value chain of telecommunication business: Infrastructure suppliers should join their efforts to shorten the time required for standardization. Spectrum regulation bodies should intensify their attempts of world-wide harmonization in spectrum regulation and should establish global cognitive radio certification institutions. Wireless network operators should closely monitor the advances in cognitive radio and should have elaborated an adequate adaption of their business strategy.

May this book be informative and stimulating to you.

Lars Berlemann & Stefan Mangold
Bonn & Berne

Acknowledgments

Writing a book creates costs. Throughout the last two years, our environments have suffered from the intense research work that we often conducted during valuable leisure times. Our friends and families often declared us foolish, in particular when vacation times were again and again used to sit in front of laptops to analyze our newest research activities and results. Working in management for great companies as well as being lecturers at universities did not really help us in finding intervals to edit our texts. It took us, our close friends and families, considerable effort and support from many sides finally to have this book finished. We apologize to all of you, dear friends and families, and thank all of you for unlimited patience and encouragement.

We thank the editors from Wiley, in particular Sarah Tilley and Sarah Hinton for their great support, many motivating discussions, professional relationship, and the many creative ways of putting friendly pressure on us.

Many thanks to our fellow colleagues from the teams at ComNets, Swisscom, Deutsche Telekom, and Philips Research for fruitful collaboration over many years.

Thanks to Deutsche Telekom and Swisscom for providing us with the opportunity to work in inspiring and innovative environments where every day new fascinating things can be learned about telecommunications markets.

We further wish to specially thank our former PhD advisor, Professor Dr.-Ing. Bernhard H. Walke of ComNets of RWTH Aachen University for his continuous support.

We both perform research on wireless communications because we really enjoy it, and because we believe in the need for the information society. We have been working together long days, at work and at home, during hikes through the Swiss Alps, during countless dinners and lunches in the wonderful restaurants of the city of Berne. The idea of writing this book was initiated by an intense discussion at Rockefeller center in New York City, a couple of years ago. We now share our research results in the exciting field of cognitive radio and dynamic spectrum access with the community, and thank everybody for her or his interest in our results.

Lars Berlemann and Stefan Mangold

Abbreviations

3G	Third generation
4G	Fourth generation
AC	Access category
ACK	Acknowledgement
AIFS	Arbitration interframe space
AIFSN	Arbitration interframe space number
AODV	*Ad hoc* on demand distance vector
AP	Access point
ARIB	Association of radio industries and businesses
B3G	Beyond third generation
BAR	Block acknowledgement request
BCH	Broadcast channel
BCH	Burst control header
BNetzA	Bundesnetzagentur
BS	Base station
BSD	Berkeley Software Distribution
BSHC	Base station hybrid coordinator
BSS	Basic service set
BW	Bandwidth
BWA	Broadband wireless access
C	Cooperation
CA	Collision avoidance
CAPEX	Capital expenditure
CBP	Coexistence beacon protocol
CCA	Clear channel assessment
CCF	Common channel framework
CCHC	Central controller hybrid coordinator
CDF	Complementary cumulative distribution function
CDMA	Code division multiple access
CEPT	European Conference of Post and Telecommunications Administrations
COOP	Cooperation strategy
CPC	Cognitive pilot channel
CPE	Customer premise equipment
CPG	Conference preparatory group

CRA	Cognitive radio architecture
CS	Carrier sense
CSCC	Common spectrum coordination channel
CSMA	Carrier sense multiple access
CSI	Channel state information
CTS	Clear-to-send
CTX	Clear-to-switch
CW	Contention window
D	Defection
DARPA	Defense advanced research projects agency
DCF	Distributed coordination function
DDR	Digital dividend review
DECT	Digital enhanced cordless telecommunication
DEF	Defection strategy
DFS	Dynamic frequency selection
DIFS	DCF interframe space
DL	Downlink
DRP	Distributed reservation protocol
DRRM	Distributed radio resource management
DS	Downstream
DS	Distribution system
DSE	Dependent station enabling
DSL	Domain specific language
DSM	Distribution system medium
DSSS	Direct sequence spread spectrum
DySPAN	Dynamic access spectrum access networks
ECA	European common allocation
ECC	Electronic communications committee
ECSA	Extended channel switch announcement
EDCA	Enhanced distributed channel access
EIFS	Extended interframe space
EIRP	Equivalent isotropic radiated power
ERC	European Radiocommunications Committee
ERMES	European Radio Messaging System
ESS	Extended service set
ETSI	European Telecommunications Standards Institute
FCC	Federal Communications Commission
FCH	Frame control header
FDD	Frequency division duplex
FDMA	Frequency division multiple access
FHSS	Frequency hopping spread spectrum
FM	Frequency management
FP6	Framework program number six
FP7	Framework program number seven
FWA	Fixed wireless access
GENI	Global environment for network innovations

GI	Guard Interval
GPS	Global positioning system
HC	Hybrid coordinator
HCCA	Hybrid coordinator controlled access
HWMP	Hybrid wireless mesh protocol
IBSS	Independent basic service set
IEEE	Institute for electrical and electronics engineers
IFS	Interframe space
IMT	International mobile telecommunications
IR	Infrared
ISM	Industry, scientific and medical
ITU	International Telecommunication Union
ITU-R	ITU radio regulations
JTP	Java theorem prover
LLC	Logical link control
LTE	Long term evolution
MAC	Medium access control
MBOA	Multiband OFDM alliance
MB-OFDM	Multiband orthogonal frequency division multiplexing
MDA	Mesh deterministic access
MDAF	MDA fraction
MDAOP	MDA opportunity
MIC	Ministry of Internal Affairs and Communications
MII	Ministry of Information Industry
MIMO	Multiple input multiple output
MNA	Mesh network alliance
MPDU	Mac protocol data unit
MPHPT	Ministry of Public Management, Home Affairs, Posts and Telecommunications
MPP	Mesh portal point
MSDU	Mac service data unit
MSG	Multi stage game
NACCH	Network access and connectivity channel
NAV	Network allocation vector
NE	Nash equilibrium
NG	Next generation
NGMN	Next generation mobile networks
NTIA	National Telecommunications and Information Administration
Ofcom	Office of Communications
OFDM	Orthogonal frequency division multiplexing
OFDMA	Orthogonal frequency division multiple access
OPEX	Operational expenditure
OSI	Open system interconnection
OWL	Web ontology language
PCO	Phased coexistence operation
PHY	Physical layer

PIFS	The PCF interframe space
PSFD	Power spectral flux density
QoS	Quality-of-service
RAT	Radio access technology
RD	Reverse direction
RDG	Reverse direction grant
RegTP	Regulation for telecommunications and post
RIFS	Reduced interframe space
RKRL	Radio knowledge representation language
RMPA	Rate monotonic priority assignment
RR	Radio regulatory
RRS	Reconfigurable radio systems
RTG	Receive/transmit Transition Gap
RTS	Request-to-send
RTX	Request-to-switch
SCC	Standards Coordinating Committee
SCH	Superframe control header
SCI	Strategy comparison index
SDMA	Space division multiple access
SDR	Software defined radio
SE	Spectrum engineering
SIFS	Short interframe space
SLS	Spectrum load smoothing
SME	Station management entity
SNR	Signal to noise ratio
SON	Self-optimizing network
SSF	Spectrum sensing function
SSG	Single stage game
SUR	Spectrum usage right
TDMA	Time division multiple access
TELEC	Telecom Engineering Center
TFT	Tit for tat
TPC	Transmit power control
TTG	Transmit/receive Transition Gap
TV	Television
TXOP	Transmission opportunity
UCS	Urgent coexistence situation
UK	United Kingdom
UL	Uplink
UML	Universal Modeling Language
UMTS	Universal mobile telecommunication system
U-NII	Unlicensed national information infrastructure
UPCS	Unlicensed pcs
US	Upstream
UTRA	UMTS terrestrial radio access
UWB	Ultra wideband

VoIP	Voice over Internet Protocol
Wi-Fi	Wireless fidelity
WiMAX	Worldwide interoperability for microwave access
WLAN	Wireless local area network
WMN	Wireless mesh network
WPAN	Wireless personal area network
WRAN	Wireless regional area network
WRC	World Radio Conference
WRC07	World Radio Conference 2007
WWI	Wireless world research initiative
XG	Next generation communication
XML	Extendable markup language
XSD	XML schema datatypes

1

Introduction

Exciting new feature-rich, interactive, and high bit-rate multimedia services of Third Generation (3G) cellular radio systems have been promised in the past. Benefits for subscribers and increased revenues for service providers and network operators have been expected. However, the wireless research community has perceived the limitations of the existing systems in terms of user throughput and cost of operation. Consequently, research and development efforts have been initiated towards Next Generation (NG) systems that are also referred to as Beyond Third Generation (B3G) or Fourth Generation (4G) radio systems. Such future systems are expected to allow subscribers transparently to access broadband multimedia services via multiple wireless and fixed-line access networks as if they were connected via broadband modems to the Internet.

The increasing demands for wireless communication in consumer electronics applications, and personal high-data-rate networks indicate a promising commercial potential. Throughput, reliability, service quality, and the ever-present availability of wireless services are more and more demanded. The number of devices based on multiple wireless standards and technologies will therefore substantially grow in the future – exciting progress but new problems will be created with these increasingly widespread wireless communications. These problems are the limited availability of radio spectrum and the difficult spectrum coexistence of dissimilar radio systems in a shared spectrum. Until today, such problems could be neglected to a great extent because network operators have usually enjoyed the privilege of exclusive access to their parts of the radio spectrum. We are, however, now at a stage where the identified problems have to be addressed to enable further growth of these promising markets and to found a substantial basis for our future information society.

1.1 Access to Radio Spectrum

Today, access to radio spectrum is difficult as it is restricted by a radio regulatory regime that emerged over the last one hundred years. Large parts of the radio spectrum are allocated to licensed radio services in a way that is referred to as command-and-control. Open access to most of the radio spectrum is only permitted with very low transmission powers, in

Cognitive Radio and Dynamic Spectrum Access   L. Berlemann and S. Mangold
© 2009 John Wiley & Sons, Ltd

a so-called underlay sharing approach such as that used, for example, used by Ultra Wideband (UWB). The overlay sharing approach, i.e. the free access to an open spectrum, is generally not permitted.

Only some small fractions of the radio spectrum, the unlicensed frequency bands, are more or less openly available. The fraction of a radio spectrum declared as unlicensed is very small, and new unlicensed spectrum will not be available soon, as regulatory changes from licensed to unlicensed spectrum are difficult and take a long time. Changing the status of a licensed radio spectrum can be perilous and painfully slow. It takes a concerted effort between government regulatory agencies, technology developers, and service providers to achieve efficient and timely deployment.

Unlicensed spectrum is a small fraction of the entire radio spectrum. Excitingly, over the past decades, this approach has led, nevertheless, to a wide variety of new wireless standards, technologies, and services, among them the popular IEEE 802.11 Wireless Local Area Networks (WLANs), Wi-Fi, and Bluetooth for IEEE 802.15 Wireless Personal Area Networks (WPANs). The demonstrated commercial success of wireless applications operating in unlicensed spectrum, and the many radio systems utilizing this fraction of the radio spectrum, indicate that it may be helpful to change the existing radio regulatory regime towards a more flexible, open spectrum access.

The limitation and delays in spectrum access form the restricting bottleneck that slows down the development of new radio services. These radio services can substantially improve health, safety, work environment, education of people, and quality of leisure time. The expected growth of the number of radio devices based on multiple wireless standards and technologies may be delayed with the existing limitations.

1.2 Artificial Spectrum Scarcity from Unexploited Frequencies

The radio spectrum is a finite resource. With the term 'radio spectrum', electromagnetic frequencies between 3 kHz and 300 GHz are referred to. Figure 1.1 illustrates the range of frequencies that are commonly regarded as radio spectrum. Most of today's radio communications systems require rigorous protection against interference from other radio systems. Nowadays, such protection from interference is guaranteed in licensing radio spectrum for exclusive usage. Most of the radio spectrum is therefore licensed to traditional communications systems and services as indicated in Figure 1.2. However, with such an approach, spectrum resources are sometimes wasted for various reasons.

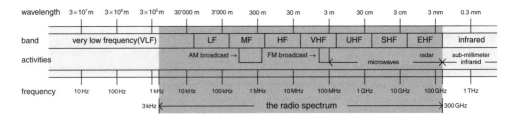

Figure 1.1   The radio spectrum refers to frequency between 3 kHz and 300 GHz

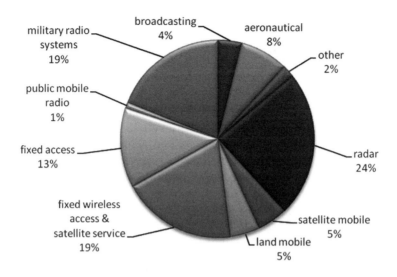

Figure 1.2 Radio spectrum is licensed to traditional communication systems

Firstly, any economic failure of licensed radio services and systems may lead to unused spectrum. For example, at the time of writing, WiMAX appears to be commercially unsuccessful. WiMAX spectrum has been allocated and licensed in many countries, but appears not to be utilized at all by nationwide network operators. WiMAX spectrum is hence considered wasted for the time being.

Secondly, public safety and military radio systems require spectrum for occasional operation, which leads to an additional amount of often unused spectrum.

Thirdly, technological progress in communications results in the improvement of the spectrum1 efficiency of existing licensed communications systems like, for example, the digitalization of television (TV) broadcasting, so that less spectrum is required to provide the same service.

As a result of all these observable trends, large parts of the spectrum are currently used inefficiently. Consequently, the traditional regulation of spectrum requires a fundamental rethinking in order to avoid waste of spectrum and, hopefully, to increase public welfare. The existing radio regulatory regime is, however, too complex to handle the increasingly dynamic nature of emerging wireless applications. This is one of the reasons why, paradoxically, $90 \ldots 95\,\%$ of the licensed radio spectrum is not in use at any location at any given time. As a result, we waste precious spectrum. What is often called 'spectrum scarcity' and 'limited radio resources' is really an artificial result of the way spectrum is regulated.

Even more problematic, the demand for additional spectrum is growing faster than the technology is able to increase spectrum efficiency, although latest research illustrated tremendous success in increasing spectrum efficiency and capacity in radio communications. As consequence, costs increase due to higher complexity for squeezing maximum data rates out of few spectrum. Multiple Input Multiple Output (MIMO) and Space Division Multiple Access (SDMA) are just two examples of the recent advances in communication technology.

1.3 Cognitive Radio and Dynamic Spectrum Access as Solution

Dynamic spectrum access refers to the time-varying, flexible usage of parts of the radio spectrum under consideration of regulatory and technical restrictions.

Cognitive radios together with dynamic spectrum access attempt to overcome the described problems. Cognitive radio is not only a new radio technology, it also includes a revolutionary change in how the radio spectrum is regulated. New radios designed for efficiently using shared spectrum and not causing at the same time significant harmful interference to incumbent (primary, license holding) radio systems are referred to as 'cognitive radios' (Mitola, 1995, 2000).

Cognitive radios are radio systems that autonomously coordinate the usage of spectrum. They identify radio spectrum when it is unused by the incumbent radio system and use this spectrum in an intelligent way based on spectrum observation. Such unused radio spectrum is called 'spectrum opportunity,' also referred to as 'white space.'

Current existing radio systems are not fully designed for mutual coordination and information exchange. New radio systems need intelligent capabilities to support Quality-of-Service (QoS) when sharing spectrum with dissimilar radio systems and in the presence of systems which have no coexistence capabilities. It is natural to foresee an evolutionary step from existing standards towards extended standards comprising the cognitive radio features. Future cognitive radio systems and networks will then autonomously coordinate themselves to support QoS in scenarios where spectrum is shared, i.e. in the presence of other, possibly competing, radio systems.

The cognitive radio technology, and the ideas discussed here, are inspired by the Defense Advanced Research Projects Agency (DARPA) Next Generation Communication (XG) program. Many other trends in research and industry are emerging. For example, a new standard IEEE 802.22 is currently being defined. This standard is a modification of IEEE 802.16 and will enable what is referred to as Wireless Regional Area Networks (WRANs). Such networks operate in rural regions in the unused VHF/UHF bands, and include mechanisms to protect incumbent radio systems (such as terrestrial TV broadcasting) from harmful interference.

1.4 This Book

Within this book, we summarize and discuss the main concepts behind radio regulation, dynamic spectrum access, and cognitive radio. In the next chapter, the way radio spectrum is handled today is discussed. We also introduce spectrum measurements that were taken by our team to illustrate the inefficiencies in today's spectrum usage. In Chapter 3, already known and in part already used approaches to improve the spectrum utilization are summarized. The following chapter, Chapter 4, lists institutions and standardization approaches in academic research and industry, all of which will lead to commercial exploitation of dynamic spectrum access and cognitive radio. This list includes a thorough discussion of IEEE 802.11n for MIMO and IEEE 802.11s for mesh networks, because both standards enable greater flexibility and spectrum efficiency in WLANs, which is important for cognitive radio solutions. We propose a number of enablers for horizontal spectrum sharing in Chapter 5 and for vertical spectrum sharing in Chapter 6 that may help to establish the concepts of dynamic spectrum access and cognitive radio. Among others, we discuss in Chapter 5 unlicensed WiMAX and coexistence with Wi-Fi, spectrum policies, and solution concepts for coexistence approaches derived from game theory. In Chapter 6 a new concept for Wi-Fi operation in the Frequency Division Duplex (FDD) spectrum allocated for WiMAX or UMTS, a dual beaconing concept for operator-assisted cognitive radio and finally our concept of spectrum load smoothing derived from 'waterfilling' are discussed. In Chapter 7, we provide an outlook on our vision about what we believe a true cognitive radio will be composed of and its potential impact on the telecommunications business. The book ends with concluding remarks in the final chapter, Chapter 8, with some additional information on our tools in the Appendix.

2

Radio Spectrum Today – Regulation and Spectrum Usage

Regulating radio spectrum is valuable and is needed for economic, societal, and technological reasons. To regulate radio spectrum in an efficient and harmonized way will be useful during the decades to come. In this chapter, we discuss the established methods, the governing institutions, and the terminology used in state-of-the-art radio spectrum regulation. Regulatory bodies such as government agencies execute spectrum regulation by determining how particular bands of spectrum could be used, making rights available to licensees or so-called unlicensed users, and finally by defining the associated spectrum licensing regime, i.e. regulatory rules constraining the access to this spectrum. The regulators' strategic decision making usually targets increasing public welfare, safety, societal benefits, and reflects public interest.

Radio spectrum is a public resource used for a wide variety of services. Utilizing radio spectrum usually means emitting electromagnetic radiation at radio frequencies (between 30 kHz and 300 GHz). This may be needed for the intended service, but may create undesirable effects for other services. For example, when communicating using commercial radio devices, unacceptable interference could be caused for dissimilar, neighboring, radio receivers that may be sharing the same or similar spectrum.

In astronomy, radio telescopes are used to observe space objects at distant parts of the Universe. The use of similar radio frequencies for other services close to the telescopes may create interference and undesirable effects for such scientific astronomical observations. As more and more spectrum is used for commercial applications, astronomical observations from Earth become more difficult. Radio regulators constantly negotiate at national and international levels a balance between the parties of such conflicting interests.

Until today, radio spectrum regulation is used to mitigate all such undesirable effects and is, therefore, considered to be inevitably required in order to enable reliable and efficient spectrum usage. We will see later why this may change with the establishment of cognitive radio and dynamic spectrum access.

Cognitive Radio and Dynamic Spectrum Access   L. Berlemann and S. Mangold
© 2009 John Wiley & Sons, Ltd

Some parts of the following sections have already been discussed elsewhere (Walke *et al.*, 2006).

2.1 History and Terminology

In April 1912, the passenger liner *Titanic*, at that time the largest passenger steamship in the world, hit an iceberg during its first voyage ever, and sank within only a couple of hours. This tragedy was not only one of the worst maritime disasters in history, but was also used as an early example of the need to introduce radio spectrum regulation. Many passengers were rescued because other ships received the radioed emergency signals, but lives were also lost because the radio receiver on board the nearest ship was not operating during the night of the tragedy. This event confirmed the importance of radio communication for public safety, and the dangers of unreliable communication (Lucky, 2001).

Over time, the way radio spectrum is used became subject of international debate, and was soon regulated by national authorities. In the US, regulation of radio spectrum was organized similarly to the economic regulation of railroads: the so-called legal Communications Act of 1934 and its predecessors provided means to control a monopoly power when a market is served by a single provider. The objective of this Act was to guarantee services 'to the public on a nondiscriminatory basis to fair and reasonable prices and conditions' (May *et al.*, 2005). This approach was adopted for radio spectrum regulation, and since the beginning of the Twentieth century, communications services like telephony, radio and terrestrial broadcast television, have been regulated according to this model.

2.1.1 The Four Basic Approaches for Radio Spectrum Regulation

The regulation of radio spectrum can be differentiated into four approaches described below. A milestone report of the US Federal Communications Commission's (FCC) Spectrum Policy Task Force (FCC, 2002) defines spectrum regulatory mechanisms in a similar way. The FCC is the regulatory authority in the US regulating commercial radio services (see Section 2.2.7).

2.1.1.1 First Approach: Licensed Spectrum for Exclusive Usage

Licensed spectrum for exclusive usage (FCC: exclusive use model) is enforced and protected through the regulator. The licensee has exclusive and transferable flexible usage rights for a specific spectrum. Frequency bands sold for use in the Universal Mobile Telecommunication System (UMTS) in Europe are an example of an exclusive usage right for a licensed spectrum.

2.1.1.2 Second Approach: Licensed Spectrum for Shared Usage

Licensed spectrum for shared usage (FCC: command-and-control model) is restricted to a specific technology. The spectrum assigned to Digital Enhanced Cordless Telecommunication (DECT)[1] in Europe is an example of this model. Another example is the spectrum used for

---

[1] Whereas DECT operates in the 2.4 GHz unlicensed spectrum in the USA. Similarly to Bluetooth and Wireless LAN, in Europe the frequency band 1880 MHz–1900 MHz is dedicated to DECT systems.

public safety services, where protection is essential for reliable service provisioning. The secondary usage of under-utilized licensed spectrum through intelligent radio systems, which will be discussed later in this book, is different to this type of sharing.

2.1.1.3 Third Approach: Unlicensed Spectrum

Unlicensed spectrum (FCC: commons model or open access) is available to all radio systems operating in conformity with regulated technical etiquettes or standards, like the Unlicensed National Information Infrastructure (U-NII) bands in the US at 5 GHz, or the Part 15 regulatory rules, see Section 2.3.2.3. Especially in Europe, the phrase 'license-exempt spectrum' is used as synonym for 'unlicensed spectrum.' In unlicensed spectrum, no right for protection from interference exists.

2.1.1.4 Fourth Approach: Open Spectrum

Open spectrum (beyond the scope of the FCC definitions above) allows anyone to access any range of the spectrum without any restriction, under a minimum set of rules from technical standards or etiquettes that are required for sharing spectrum. The term 'open access' has a different meaning in the context of the spectrum auctions at 700 MHz than open spectrum. 'Open access' refers in this context to the freedom for the customer to use any device and software in the spectrum at 700 MHz.

2.1.1.5 The FCC terminology

The FCC's task force report (FCC, 2002) classifies the assignment of spectrum rights into the 'exclusive use' model, the 'command-and-control' model, and the 'commons' or 'open access' model. The 'command-and-control' is currently the most often used regulation model and refers to the 'licensed spectrum for shared usage' and 'unlicensed spectrum' in the four approaches described above.

2.1.2 Guiding Principles

A magazine paper by Peha provides a very useful taxonomy of guidelines and objectives for spectrum regulation (Peha, 2000). According to Peha, radio spectrum regulation influences the development of spectrum access standards to balance the six objectives given below.

First, adequate Quality-of-Service (QoS) should be feasible to all radios depending on the supported applications. Second, no radio should be blocked from spectrum access and from transmission for extended time durations. Third, spectrum regulation and standards must not slow down innovations in the economically successful and rapidly changing wireless communication markets. Fourth, the available spectrum should be used efficiently, including spatial reuse of spectrum and solving the 'tragedy of commons' that will be discussed later in Section 2.3. Fifth, spectrum should be used in a dynamically adaptive way, taking the local communication environment including spectrum usage policies into account. And finally, sixth, the costs of commercial radio devices should not increase significantly through techniques mandated by radio regulation.

2.2 Institutions that Regulate Radio Spectrum

Licensing spectrum for exclusive or shared usage, and declaring spectrum as unlicensed or even open spectrum, is organized by national and international institutions that are usually referred to as 'regulators.' Some of these regulators are briefly introduced in this section, together with their basic strategic approaches. The guidelines from Peha (2000) that are discussed in the previous section can be found in many of the recent statements of nearly all regulators discussed below.

2.2.1 International Telecommunication Union, ITU

The process of licensing is often referred to as 'frequency allocation,' or, 'spectrum allocation.' Globally, for all regions of the world, frequency allocation processes are harmonized with the help of the Radiocommunication Sector of the International Telecommunication Union (ITU-R). The ITU is the United Nations agency for information and communications technologies, and ITU-R attempts to ensure the rational, equitable, efficient and economical use of the radio-frequency spectrum by all radio communications services. ITU agreements on spectrum allocation are set out in the ITU Radio Regulations (ITU RR), which have treaty status. The ITU RR regulate the use of radio spectrum internationally and form the global framework for regional and national planning. Nations remain however sovereign in their use of radio spectrum in their own territory, and Article 48 of the ITU constitution states that ITU members may retain their freedom with regard to military radio use. The ITU RR are regularly revised through World Radio Conferences (WRCs) that take place every 2–3 years. At the WRCs, efforts are made at international level to cope with the many emerging services, with conflicting interests, and business models, in an attempt to modernize spectrum usage.

At the time of writing, the most recent WRC took place in Geneva in 2007. At WRC07, the main discussion concerned the harmonization of the radio spectrum for global use by International Mobile Telecommunications (IMT-2000 and IMT-Advanced) systems. These IMT technologies consist of existing and future mobile radio network technologies like Long Term Evolution (LTE), UMTS, and WiMAX. In particular, the following spectrum was identified as candidate spectrum for IMT systems:

20 MHz in the band 450–470 MHz (worldwide);
72 MHz in the band 790–862 MHz for Europe and Asia;
108 MHz in the band 698–806 MHz for the Americas;
100 MHz in the band 2.3–2.4 GHz (worldwide);
200 MHz in the band 3.4–3.6 GHz (agreed between many nations).

This would mean, in total, up to 392 MHz in Europe and up to 428 MHz in the Americas of newly allocated spectrum for IMT systems. Additionally, coordination with national licenses (i.e. TV broadcasting) will be required.

2.2.2 Europe

At the European level, spectrum regulation is conducted through the Electronic Communications Committee (ECC) of the European Conference of Post and Telecommunications Administrations (CEPT) that currently, in 2008, has 46 member countries. Frequency allocation issues are handled through the Frequency Management Working Group (FM) and there are also committees coordinating Radio Regulatory (RR) and Spectrum Engineering (SE) issues. The ECC structure was changed in October 2001 when the responsibilities for the radio and telecommunications sides of the CEPT – previously handled separately – were combined in the new ECC.

Each of the ECC's three main working groups (FM, RR and SE) comprises a number of project teams that are charged with detailed examination of specific areas. The ECC produces reports, recommendations and decisions on spectrum usage. When implemented by member countries, these decisions form the basis for European harmonization of spectrum usage at the allocation level. Unlike ECC recommendations and decisions, directives are binding on member states of the European Union. In 1992 it was agreed that European spectrum harmonization should be carried out through CEPT and that EU member states should commit themselves to full participation in the development and implementation of ECC decisions.

The ECC coordinates long-term spectrum planning in Europe and has produced the European Common Allocation Table (ECA table), a harmonized frequency table for Europe, that describes the entire usable radio spectrum on nearly 300 pages. The ECC also provides the forum for coordinating European preparations for WRCs through its Conference Preparatory Group (CPG).

The EU also implemented several spectrum harmonization measures. In the early 1990s, three frequency harmonization directives were executed: on GSM, DECT and the European Radio Messaging System (ERMES).

In recent decades, Europe privatized former state owned telecommunications companies to overcome market entry barriers and to harmonize the member state's law. The *'New Regulatory Framework'* (EU, 2002b) is aimed at establishing conditions for fair and efficient competition in eliminating sector specific regulation (EU, 2002b). The regulatory framework assumes a 'significant market power' (EU, 2002b) as the basis for regulation in most cases. The framework sees a need for minimum regulation to enable the interconnection of all public communications networks and providers aiming at a single interoperable network. Under this EU framework, the finding of a significant market power permits the individual spectrum regulation of national regulators belonging to the individual member states. The *'radio spectrum decision'* (EU, 2002a) aims at the harmonization of conditions for spectrum usage considering availability and efficient use of the radio spectrum. This decision was seen as being necessary for establishing and functioning of internal competitive markets for electronic communications, transport, and research and development.

The European Commission advocates, in a strategy paper (EC, 2007), a flexible spectrum management and states that from a market-based spectrum management combined with flexible spectrum usage rights, it expects a net gain of €8–9 billion per year across Europe (EC, 2007). In this strategy paper, the freeing up of digital broadcasting frequencies is identified as a 'high priority case' (EC, 2007) for introducing flexible spectrum usage. The target date for introducing flexible spectrum management with individual usage rights as part of a pan-European spectrum regulation is set at 2010.

2.2.3 Germany

What the ECC decides is usually realized by the CEPT member countries such as Germany. In Germany the Bundesnetzagentur (BNetzA) formerly referred to as the Regulatory Authority for Telecommunications and Posts (RegTP) is responsible for spectrum regulation.

The emphasis of Germany's frequency regulation in the next years was published in 2005 (Bundesnetzagentur, 2005). The status quo and strategies for the future are introduced for many frequency bands. This includes, for example, the global harmonization of frequencies assigned to Wireless Local Area Networks (WLANs) and a frequency refarming. The future of UMTS, GSM, and DECT frequency bands, frequency trading and strategies for Ultra Wideband (UWB) applications were announced to be included in a future revision of this publication.

The published BNetzA objectives can be compared to what the European Commission published as its future strategy (EC, 2007). BNetzA also targets at full transparency of future spectrum regulation and aims at giving guidelines for future research, innovation and investment decisions towards such targets. Similarly to the ambitious strategy of the European Commission (EC, 2007), it remains to be shown what part of the published BNetzA targets will finally be implemented.

2.2.4 United Kingdom

The Office of Communications (Ofcom) is the regulation authority in the United Kingdom (UK) and is responsible for the regulation, management, licensing and assignment of radio spectrum. Established in 2003, Ofcom operates with a process of regular consultations: initial statements are published and the public is asked to review and comment on them.

Ofcom's vision of spectrum regulation has three main pillars (Ofcom, 2004). Firstly, spectrum should be free of technology and usage constraints as far as possible. Policy constraints should only be used where they can be justified. Secondly, it should be simple and transparent for license holders to change the ownership and use of spectrum. Thirdly, rights of spectrum users should be clearly defined and users should feel comfortable that they will not be changed without good cause. Ofcom intends to archive this (Ofcom, 2004) by:

> 'providing spectrum for license-exempt (unlicensed) use as needed, but [. . .] Ofcom's . . . current estimates are that little additional spectrum will be needed in the foreseeable future, growing to 7 % of the total spectrum;
> allowing market forces to prevail through the implementation of trading and liberalization where possible. We believe we can fully implement these policies in around 72 % of the spectrum; and continuing to manage the remaining 21 % of the spectrum using current approaches'.

In its Digital Dividend Review (DDR) (Ofcom, 2008), Ofcom recently declared that 128 MHz of spectrum (at frequencies below 854 MHz) will be cleared of existing uses UK-wide by 2012. It is intended to let the market decide the future usage of this spectrum by auctioning tradeable Spectrum Usage Rights (SURs). The SURs will be tailored to the different transmission network types that are likely to be deployed in the spectrum.

2.2.5 Japan

In 2004, the Ministry of Internal Affairs and Communications (MIC) changed its name from Ministry of Public Management, Home Affairs, Posts and Telecommunications (MPHPT) to MIC. MIC is responsible for planning, designing and promoting general policies for using communications devices and is in charge of frequency allocation. The Association of Radio Industries and Businesses (ARIB) is designated by the MIC as 'the Center for Promotion of Efficient Use of Radio Spectrum' and 'the Designated Frequency Change Support Agency' (ARIB, 2008). ARIB conducts studies, establishes standards and provides consultation services for radio spectrum coordination and cooperates with overseas organizations. Currently ARIB has 277 members from the field of telecommunications, broadcasting services, and research, development and manufacturing of radio equipment. The Telecom Engineering Center (TELEC) in Japan provides services related to technical regulation conformity certification under the designation of the MIC.

The MIC recently identified a bandwidth of at least 1.5 GHz below 6 GHz by 2013 as a requirement for future needs in its radio policy vision (MIC, 2003). There, a review of frequency assignments, establishment of rules enabling a fast and simple reallocation of spectrum, and a reconsideration of the licensing system for using spectrum, has been identified as cornerstones of future spectrum management. The MIC is establishing a legal basis for a compensation system in spectrum reallocation. The economic benefits for new spectrum users are used to collect an additional usage fee in order to compensate for the economic losses of the incumbent radio system. The freeing of the 100-MHz band from 4.9 to 5.0 GHz for outdoor WLAN in metropolitan areas from being licensed for fixed microwave station for telecommunication business by 2005, is a first example of implementing such a compensation system.

2.2.6 PR China

In the Peoples' Republic of China, radio spectrum is regulated on two levels: the Ministry of Information Industry (MII) and its radio regulatory department are responsible for nationwide regulation, while local radio regulation institutions realize state regulation and administer regional regulation in the various provinces. Additionally, there is the Army Radio Regulatory Commission in China, which regulates spectrum usage for military systems. The MII is technically supported by the China Radio Monitoring Center, which supervises spectrum usage by looking for interference and unauthorized radio stations.

Spectrum regulation in China is restrictive and driven by interests to protect national economic markets. For operation of telecommunications equipment, the MII issues so-called network access licenses. Without such a license operation is not allowed. In recent years, the MII has not followed international efforts in harmonizing standards and regulation, favoring individual solutions. There was, for example, an attempt in 2003 for all radio devices based on the IEEE 802.11 standard for WLAN to require a dedicated Chinese encryption standard.

Nevertheless, it is known that the Chinese legislation currently separates its administrative and enterprise functions related to spectrum to enable competition and to control any monopoly.

2.2.7 United States of America

In the USA, the responsibility for spectrum regulation is shared between the FCC and the National Telecommunications and Information Administration (NTIA). The NTIA is responsible for the spectrum used by the government and the FCC is responsible for spectrum used for non-government purposes.

The Spectrum Policy Task Group of the FCC has identified three approaches to the improvement of spectrum utilization (FCC, 2002): (i) improve access in space, time and frequency domain; (ii) enable flexible regulation in permitting controlled access to licensed spectrum, and (iii) stimulate efficient spectrum usage through policies. Therefore, FCC is rethinking its licensing policy.

On the other hand, the FCC is initiating multiple approaches to liberalization of spectrum usage. Thereby, market orientation and flexibility are the target for providing opportunities for new technologies and stimulating investments. The additional unlicensed spectrum at 5 GHz (FCC, 2003a), the new unlicensed spectrum at 3.7 GHz (FCC, 2004b), and the opportunistic usage of TV bands (FCC, 2004c) are examples of this.

The opportunistic usage of TV bands, i.e. spectrum that is currently used exclusively for terrestrial television broadcasting, will be discussed later in this book in more detail. The basic principle is illustrated in Figure 3.6 (page 36). Additionally, the FCC is also focusing on cognitive radio technologies for flexible and reliable spectrum usage (FCC, 2003c, 2005b) and the Spectrum Policy Task Group of the FCC introduced the interference temperature concept for underlay spectrum sharing (FCC, 2003b). We will discuss all these important concepts later in this book.

2.3 Licensed and Unlicensed Spectrum

Large parts of the radio spectrum are allocated to licensed radio services in a way that is often referred to as exclusive usage plus command-and-control. Licensing spectrum mainly covers the exclusive access to spectrum and spectrum sharing within the licensed spectrum, which are the first two approaches described in Section 2.1.1.

In the case of exclusive spectrum usage (first approach), a license holder typically pays a fee for this privilege. Exclusive access rights have the advantage of preventing potential interference from dissimilar radio systems, which otherwise would imply dangers to reliability and thus commercially exploitable communication services. In the case of spectrum scarcity, licensed spectrum is highly valuable, usually leading to economic profits for the licensee, as consumers need to pay to use it. Having commercial implications, spectrum licenses are typically tightly bonded to requirements that must be fulfilled by the licensee, for example, a clearly required specific transmission technology is allowed in this spectrum, or a certain percentage of population is to be reached by the network within a defined time. The auctions in the countries of Europe for licensing spectrum to 3G systems in 2000 are an example of this.

Today's often-used licensing model is to license spectrum for shared usage restricted to a specific technology (the second approach in our list in Section 2.1.1). Emission parameters like maximum transmission power and interference to neighboring frequencies such as out of band emissions are restricted by the regulators. Regulation takes care of protection against interference and sometimes of the support of rudimentary coexistence capabilities such as dynamic channel selection in DECT, which became mandatory and part of the DECT standard.

2.3.1 The Disadvantages of Spectrum Licensing

Licensing spectrum takes time and is very difficult, therefore licensing is also very expensive. The licensing process constrains innovation as it is a difficult barrier to overcome when introducing new technologies. The inflexibility of exclusive usage rights from licensing spectrum leads to inefficient spectrum utilization, as the license prohibits the usage of the spectrum if it is underutilized or even unused by the license holder. In 1959, R. H. Coase[2] had already criticized spectrum allocations by governments in general and spectrum licensing in the 'public interest' (Coase, 1959). Coase suggested the issuing of clearly defined rights of spectrum ownership as a more efficient method of allocating spectrum to users. Coase observed that 'government control' (Coase, 1960) is not required for economic development of spectrum usage but actually constrains it. The arguments led Coase to formulating a general theorem that claims that well-defined property rights and moderate trading costs for these rights lead to efficient allocation of resources (Coase, 1960). In literature, Coase is therefore often referred to as the first innovator requiring a spectrum policy reform.

Another problem with licensing is the duration of a license. Typically licenses expire after a decade, but can be renewed. Temporal licenses give the regulator the opportunity of intervening if the spectrum is under-utilized or wasted. The regulator is able to answer market demands in shifting, extending or reissuing licenses, and can thus accelerate the introduction of new technologies. The danger of temporal licenses on the other hand, is that uncertainty about future regulatory decisions may hold back investments (Peha, 1998).

Licenses are often issued based on auctions and the advantage or disadvantage of this procedure has been discussed in economic science for many years. Auctions for selling spectrum access rights (referred to as trading at the primary market as introduced below) are primarily a revenue tool for governments and do not reflect social values. An often voiced criticism is that a one-time payment from a spectrum auction is only used by governments to pay for annual expenditure, instead of reinvesting in economic growth. For an excellent discussion on auctions of spectrum licenses see the literature (Hazlett, 1998; Noam, 1998).

2.3.2 Unlicensed Spectrum as an Alternative

The availability of a adequate amount of radio spectrum is crucial for the economic success of future radio communications systems. The commercial success of wireless technologies operating in unlicensed frequency bands, especially of the IEEE 802.11 for WLANs, demonstrates the economic advantage of unlicensed spectrum. Unlicensed spectrum is available to all users but is still strictly regulated. An unlimited number of users are sharing the same unlicensed spectrum, in a way often called 'best-effort.' Spectrum usage is allowed for radio devices that satisfy certain technical rules or standards that mitigate potential interference resulting from transmission powers or advanced coexistence capabilities. The usage rights of unlicensed spectrum are flexible and no dedicated spectrum access method is specified or required.

The term 'license-exempt' is a synonym for 'unlicensed.' There are more than ten unlicensed frequency bands, but the following two are considered to be important for our discussion: The Industrial, Scientific and Medical (ISM) bands from 2400–2483.5 MHz and the frequency band between (and below) 5 GHz and 6 GHz. In the USA, the frequency band at 5 GHz is referred to as the Unlicensed National Information Infrastructure (U-NII) band. ISM and U-NII hence both refer to unlicensed spectrum. While the regulatory restrictions for the ISM bands are the same throughout different regulatory domains in the world, spectrum

---

[2] Ronald H. Coase (*1910), economist, Nobel Prize in economic science 1991.

regulation of unlicensed operation in the 5-GHz band differs essentially when comparing Europe, US and Japan. This is indicated in Figure 2.1(a). Whereas any radio system, even microwave ovens, is permitted to operate in the ISM band, only communications systems are permitted to operate in U-NII bands.

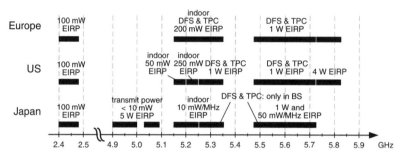

(a) Some typical regulatory restrictions for prominent unlicensed operation

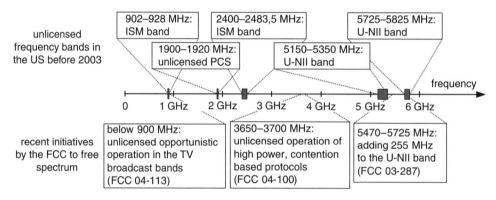

(b) Unlicensed frequency bands in the United States of America

Figure 2.1 (a) Frequency bands and regulatory restrictions of unlicensed operation in Europe, the USA, and in Japan, and (b) available spectrum for unlicensed operation in the USA. DFS = Dynamic Frequency Selection, TPC = Transmitter Power Control, EIRP = Effective Isotropic Radiated Power. Reproduced with permission from L. Berlemann, C. Hoymann, G. R. Hiertz, and B. H. Walke, 'Unlicensed Operation of IEEE 802.16: Coexistence with 802.11(a) in Shared Frequency Bands,' in Proc. of 17th IEEE Conference on Personal, Indoor and Mobile Radio Communications, PIMRC 2006, Helsinki, Finland, 11–14 September 2006. © 2006 IEEE

2.3.2.1 Europe

A systematic overview of the regulation of short-range radio devices in Europe can be found in a publication of the European Radiocommunications Committee (ERC) (ERC, 2005). The regulations for the usage of the 2400–2483.5 MHz band are described in an earlier publication (ERC, 2001). The Equivalent Isotropic Radiated Power (EIRP) is limited there to 100 mW. Further, for Direct Sequence Spread Spectrum (DSSS), the maximum spectrum power density is limited to −20 dBW/1 MHz, and for Frequency Hopping Spread Spectrum (FHSS) the maximum spectrum power density is limited to $-10\,\text{dBW}/100\,\text{kHz}$. Operation in the 5-GHz frequency bands is regulated by the ECC (ECC, 2004). The 5150–5350 MHz band is designated for indoor usage with a mean EIRP of 200 mW, and the spectrum management concepts Dynamic Frequency Selection (DFS) together with Transmit Power Control (TPC) are additionally required for unlicensed operation above 5250 MHz. The frequencies from 5470 to 5725 MHz may be used indoors as well as outdoors and the mean EIRP is limited in this band to 1 W. The usage of DFS and TPC is mandatory.

Compared with the USA, the multitude of different regulating authorities in Europe results in a slower liberalization process of radio spectrum regulation.

2.3.2.2 United States of America

Figure 2.1(b) illustrates the status of unlicensed spectrum in the US. Besides the ISM bands at 900 MHz and 2.4 GHz, the FCC opened 20 MHz at 1.9 GHz for Unlicensed PCS (UPCS) in 1990. Additionally, FCC reserved 300 MHz in 1997, and 255 MHz at 5 GHz in 2003 for unlicensed operation. Unlike the ISM bands, usage of the U-NII bands is more restricted. There, limited coexistence capabilities like DFS and TPC, are mandatory for parts of the U-NII (FCC, 2003a).

As TV bands in the US are often under-utilized, the FCC proposed allowing unlicensed systems to have secondary usage of this spectrum (FCC, 2004c). This principle, referred to as 'vertical spectrum sharing', is discussed in detail in Section 4.4.4.

In 2004, the FCC also initiated the opening of new spectrum for wireless broadband communication in the 3650–3700 MHz range for fixed and mobile devices transmitting at higher power (FCC, 2004b). It is envisaged that multiple users could share this spectrum through the use of contention-based medium access protocols, such as IEEE 802.11, in order to minimize interference between fixed and mobile operations. New fixed and mobile stations will therefore be required to use contention-based protocols, which will reduce the possibility of interference from co-frequency operation by managing each station's access to spectrum. The FCC regards this approach as a reasonable, cost-effective method for ensuring that multiple users can access the spectrum. Apart from a few regional constrains at radar sites and frontiers, fixed stations will be allowed to operate with a peak power limit of 25 watts per 25 MHz bandwidth, and mobile stations with a peak power limit of 1 watt per 25 MHz bandwidth. For further details of the FCC's understanding of contention based protocols in the context of IEEE 802.11y see Section 4.4.4.

The licensing, service and operation provisions for this spectrum will be placed in Part 90 of the FCC's rules taking the nonexclusive nationwide nature of the spectrum into account. The status of this frequency band is currently subject to intensive lobbying and emotional discussions from different protocol factions in the wireless industry.

2.3.2.3 Part 15 Regulation

The Part 15 rules of the FCC (FCC, 2005c) describe the regulations under which a transmitter may be operated without an individual license. It also contains the technical specifications and administrative requirements of Part 15 devices. ISM low-power devices, like garage openers for example, are allowed to transmit at 1 W if using spread spectrum technologies. Three basic principles describe in general the rules of Part 15 regulation: 'Listen before talk', 'when talking, make frequent pauses and listen again' and 'don't talk too loud'. When detecting a busy channel, either another unused channel is chosen or the radio waits until the channel is idle again. These simple rules require no interoperability or information exchange between spectrum sharing devices. IEEE 802.11, being designed for operation in frequency bands subject to Part 15 regulation, realizes spectrum access corresponding to these three basic principles. An analysis of the QoS capabilities of 802.11 (Mangold, 2003) shows that with the current Part 15 regulation, a QoS support is very difficult, if not impossible, in the case of coexistence with dissimilar radio systems. An additional distributed coordination is required in order to allow QoS support.

Any accidental background noise emitted from consumer electronics, such as personal computers operating at 2–3 GHz is also restricted in the Part 15 regulations.

2.3.3 Tragedy of Commons in Unlicensed Spectrum

The commercial success of unlicensed spectrum and the many different radio communications systems that operate within such spectrum, are a strong motivation for more unlicensed spectrum in the future. Further, the QoS requirements imposed by upcoming multimedia applications will not be fulfilled with today's rules to enable coexistence of dissimilar radio systems all operating in the same unlicensed spectrum (Mangold, 2003). This again motivates us to either ask for more spectrum, or more efficient radio systems.

In case of near-field and short-distance wireless communication (personal area networks), spectrum demand usually concentrates in one single location. In such scenarios, the competition for shared spectrum is modest, because it is unlikely that multiple short-distance networks would operate in a single location at the same time. This is true for at least today's penetration. Therefore, the regulatory instrument of simply restricting radio transmission, for example limiting the emission power of radio devices, can regarded as sufficient to enable at least a minimum level of service quality for all networks.

In all other scenarios, for example WLANs, unlicensed spectrum usage is a victim of its own success, with many parties and different technologies utilizing unlicensed spectrum so that it is becoming more and more heavily used – and thus less available for all. Different radio systems operate in unlicensed spectrum (WLAN, Bluetooth, DECT and many more). However, if one radio system type were to be technically improved towards higher spectrum efficiency, for example by operating with more complicated spectrum management, this would be a gain for all the competing different radio system types. The gain in spectrum efficiency would be beneficial for all radio systems, not only the one that was improved. All radio systems would potentially communicate more as the result of the improvement – the spectrum would remain less available for all. We see immediately that such unilateral improvements are not very likely to happen: there is no direct incentive to improve one radio system type, which is considered tragic as spectrum could be used more efficiently overall. In economics, this phenomenon is referred to as the 'tragedy of commons' (Hardin, 1968).

Hazlett (2005) additionally introduces the 'tragedy of the anticommons'. Contrary to the overuse of spectrum due to missing regulation of access, the tragedy of the anticommons refers to inefficient spectrum utilization because of too restrictive regulation.

The tragedy of commons and the associated inefficient overuse of spectrum result in less investment in improved technology, and directly challenges the open-access approach.

Therefore, regulators impose restrictions such as limiting transmission powers. As a consequence, many alternative radio systems are not permitted to operate in such a spectrum, which results again in inefficient under utilization of spectrum. Hazlett is concluded that restricting the level of spectrum sharing by regulatory rules is the only way out of this tragedy. When we discuss cognitive radio, we see later that such statements are not really needed, as spectrum coordination can be realized with more flexible alternatives than restricting the amount of spectrum sharing.

How intensively is our radio spectrum used? Today, a multitude of different radio systems operates in the radio spectrum. Not only communications systems, but numerous other types of service are regulated, including Earth and deep space observations. Military systems, broadcasting, various radars for different applications, the Global Positioning System (GPS), and many other radio systems have found their place in the spectrum. The large number of existing radio systems and technologies that operate in the radio spectrum indicates that overall, the whole radio spectrum could be occupied at any given time and location. After all, there are hundreds of different radio systems and services, and in the case of a new service being launched, long debates and searches are taking place before a candidate frequency band is finally been identified for the new service.

So what is the reality in spectrum usage? How intensively is our radio spectrum utilized? Is radio spectrum completely crowded? Is it really a scarce resource that we have to protect against too many undesirable usage scenarios? We show in the following the results of a measurement campaign that we conducted in order to find the answers to these questions. Figure 2.2 illustrates the results of the measurements. Our results confirm what is already known from literature: The reality is different to what might be expected: many frequency bands are not used at all, and the spectrum appears to be under-utilized for most of the time, in most of the locations.

2.3.4 Spectrum Measurements

Figure 2.2 illustrates the results of a measurement campaign for spectrum utilization in an office area in Berne, Switzerland. Figure 2.2(a) illustrates what can be expected if no system were to operate in the radio spectrum. Figures 2.2(b) and (c) show the usage of spectrum in an office in the city of Berne, Switzerland, at different frequencies.

We performed spectrum measurements using an omnidirectional antenna and a hand-held spectrum scanner with channel sounding capability. Measurements were made at different locations in the city and in rural areas. Shown in the figure are the results from the measurements taken in our office building. Measurements were conducted always for a time interval of around 24 hours. With each repeated channel sweep, the spectrum was divided into slots of 20 kHz in which the received signal power was collected. All energy was collected; no demodulation or matching to a particular radio signal took place.

2.3.4.1 Measurements in an Isolated Environment

Figure 2.2(a) illustrates the spectrum utilization in a closed chamber, where no signal is sensed and consequently the complete spectrum is shown darker. Some light areas indicate that at some frequencies electromagnetic noise found its way into the chamber. The chamber where

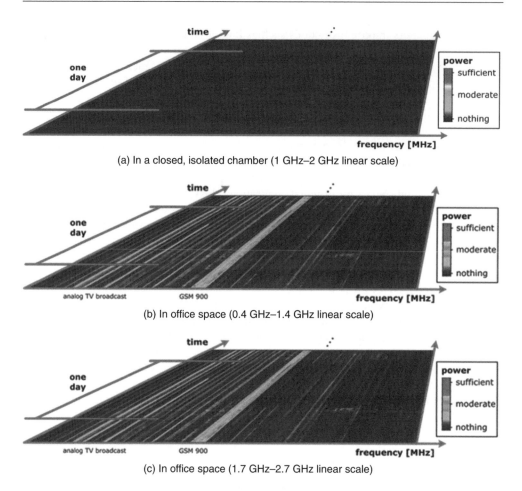

Figure 2.2 Spectrum utilization measurements over one day. Three different scenarios are shown. Darker areas indicate no use or passive receive-only usage, lighter areas indicate activities

this measurement was taken was a nuclear bunker on the third underground floor of a public office building in Berne, Switzerland. The shelter doors were left open during the measurement campaign inside the bunker, to avoid our technicians getting worried when setting up the equipment.

2.3.4.2 Measurements in an Office Environment, 400 MHz to 1.4 GHz

After the measurements in the basement, where no significant signal energy was detected, the measurement equipment was moved and set up in an exposed office building in the center of the city. In this measurement, the channel sounder swept through the spectrum from 400 MHz to 1.4 GHz. The results are illustrated in Figure 2.2(b). Analog TV broadcast and the GSM cellular radio systems at 900 MHz can be identified. TV channels do not dynamically change power or frequency over time and therefore form a predictable pattern with on–off structures of every 8 MHz, which is the channel bandwidth of analog TV in Europe. One transmission end of one TV channel during the early hours of the morning can be observed (between 00:00 and 06:00). There are channels that are switched off during the night. The measurement was performed in 2006, right before the transition towards digital TV in Switzerland, which is the reason why only analog TV channels were detected. The change towards digital TV broadcast, i.e. DVB-T, will also influence this visible usage pattern and the utilization of these frequencies.

We can see that between the active TV channels, there is unused spectrum. Because of the static nature of the analog TV channels (no changes over time), such unused spectrum is easy to detect using a radio device. Some other areas exist in the graph that are also marked darker. These areas are either used by passive radio systems or not used at all. Radio telescope exploration radars, military, and public safety applications are typical examples using such spectrum.

2.3.4.3 Measurements in Office Environment, 1.7 GHz to 2.7 GHz

The uplink and downlink of the GSM systems at 1.8 GHz, the UMTS broadcast channels at 1.9 GHz, and unlicensed WLANs at 2.4 GHz are visible in the measurement results from 1.7 GHz to 3 GHz shown in Figure 2.2(c). This measurement campaign was conducted in the same location as before. The three characteristic channels of Wireless LAN IEEE 802.11b can be identified: one channel is more intensively used because this is the channel at which the corporate WLAN was set up in the office where the measurements were taken, while the other two channels are less aggressively used.

The three UMTS broadcast channels of the three 3G operators in Switzerland are visible, but not much client activity can be seen. This lack of UMTS activity for all 3G operators in early 2006 confirms the slow take up of 3G clients at that time, after the UMTS services were started. The varying utilization of the GSM systems at 1.8 GHz indicates the busy times and day times, and shows how intensively this system is used.

2.3.4.4 Learnings

Radio spectrum is not heavily used, as it is inefficiently allocated. The measured spectrum usage is mainly static and the remaining spectrum (the darker areas) seems to be under-utilized. Future concepts that will be discussed in subsequent chapters are the introduction of greater flexibility in spectrum licensing, in order to mitigate the undesirable effect of low spectrum usage. Further, the concept of dynamic spectrum access that will be discussed later in the book makes this underutilized spectrum available for secondary usage, without interfering with the primary (incumbent) radio systems. The objective of dynamic spectrum access is to achieve a more efficient utilization of our radio spectrum.

3

Radio Spectrum Tomorrow – Dynamic Spectrum Access and Spectrum Sharing

Dynamic spectrum access and spectrum sharing are tools that provide regulators with the flexibility needed in order to achieve a more efficient spectrum usage. We saw in the previous chapter that the current understanding of regulators today is that commercial broadband and cellular networks indeed all require exclusive spectrum access to guarantee the support of QoS for their services. However, nearly all regulators already actively express their interest in exploring the path towards greater flexibility in spectrum licensing processes to reflect the apparent fast developments in the wireless communication market.

In what follows, established and newly proposed concepts and techniques for realizing a greater flexibility in spectrum regulation are described. We focus on approaches to increase the overall efficiency of spectrum utilization, referred to as dynamic spectrum access and spectrum sharing. Some of the discussed concepts are already used today in various countries for parts of the radio spectrum; other concepts are new and still at the level of academic research.

We will further introduce and make use of important expressions in this chapter: 'underlay/overlay spectrum sharing' in Section 3.1.2, and 'horizontal/vertical spectrum sharing' in Section 3.1.3. We provide a taxonomy of these concepts and enhancements based on coexistence, mutual coordination, cooperation and defection.

3.1 Spectrum Sharing and Dynamic Spectrum Access: Concepts and Terminology

Regardless of the regulatory model, flexibility and efficiency need to be reflected in spectrum access. Techniques that sense and adapt to the radio environment are, for example, essentially required in unlicensed bands and improve spectrum access through methods such as secondary markets. Spectrum sharing hence plays an important role in increasing spectrum

Cognitive Radio and Dynamic Spectrum Access L. Berlemann and S. Mangold
© 2009 John Wiley & Sons, Ltd

utilization, especially in the open spectrum context. Below, we differentiate between primary and secondary users of spectrum. Secondary users operate with a lower regulatory priority and have to defer to primary users by vacating spectrum immediately when primary users need the radio resources.

3.1.1 Spectrum Trading and Spectrum Liberalization

In the context of radio spectrum regulation, it is necessary to distinguish between 'trading,' as transfer of spectrum usage rights, and 'liberalization,' as weakening of restrictions and limitations associated with spectrum usage rights related to technologies and services.

National regulation authorities indicate their willingness to liberalize the regulation of radio spectrum. However, the concrete realization of liberalization differs when comparing the approaches and public statements of national regulation authorities.

Currently, spectrum is a scarce and valuable source. The auctions of 3G spectrum licenses in Germany and UK have shown that operators pay high values for the rights to use this spectrum. However, these high prices reflect the spirit currently existing in the wireless communication industry and at the stock markets. Economic development in wireless communication in recent years indicates that such high prices will not soon be reached again, as the return of investment in 3G spectrum takes an unexpectedly long time. This long time before return of investment challenges in general the regulatory approach of auctioning spectrum licenses, especially when taking into account that the way the 3G spectrum was licensed artificially increased the license fees, sometimes to excessive values (Valletti, 2001). The regulatory prohibition of using other spectrum with 3G technologies on the one hand, and the limited amount of spectrum available for exclusive use on the other, led to prices not reflecting the market value of spectrum – particularly because government-owned bidders participate at the 3G auctions.

Spectrum can be subdivided into quantities in many domains (e.g., frequency, time, code). Trading of spectrum refers to the transfer of spectrum usage rights for a certain quantity of spectrum, usually specified frequency bands. Compared with trading with securities, spectrum trading is performed at two markets. The initial issuing of spectrum through regulators is executed at the so-called primary market, while the transfer of spectrum usage rights between different parties can then be executed at the so-called secondary market. In contrast to licensed spectrum, spectrum trading implies that the owner of spectrum usage rights may differ from the user of spectrum. Spectrum can be subdivided into many dimensions to be resold by the license holder to different parties.

It is said that the trading with radio spectrum will increase the efficiency of spectrum usage and is therefore attractive for regulation: The spectrum owner, such as, for example, a spectrum broker, has a financial interest in promoting spectrum efficiency and innovation (Peha, 2000). Spectrum trading in the secondary market has impacts on regulation as the owner and the user of spectrum are different parties. In 2004, the FCC therefore made secondary markets for trading spectrum legal, allowing the licensee for the duration of its license to lease the rights to use spectrum (FCC, 2004a). Depending on the selected option for transferring usage rights the regulator might intervene in the pricing mechanisms in the 'public interest'.

Valletti (2001) recommends spectrum trading based on fully flexible transferable rights, which may be traded. The regulator intervenes in the case of market failures on the assumption that the market will usually solve problems on its own. The enforcement of antitrust laws and arbitrating in case of disputes are the task of the regulator. Additionally, Valletti (2001) suggests that the regulator licenses low power equipment for open access spectrum.

Spectrum usage rights can be transferred in different ways, as known from other assets (economic goods):

Lease: The right to use spectrum is temporarily transferred while the ownership remains with the license owner. The license holder can define their own rules additionally to regulation for using their spectrum. This option includes, for example, spectrum brokers that lease spectrum in determining prices on the basis of auctions.

Sale: The ownership of usage rights is permanently transferred.

Options and futures: The right to access spectrum at a certain future point of time for a predefined duration is transferred for a pre-agreed price. Compared with the financial markets, trading with options and features can lead to complex constructs. Financial contracts will emerge to provide security against movement of spectrum prices, i.e. financial payments for reallocating risks.

A recent study of Analysys Consulting (2004) recommends one common overall approach to spectrum trading and liberalization of spectrum usage. A detailed implementation of the spectrum trading framework is demanded, which refers to the creation of tradable rights and the establishment of an adequate forum for trading/transfer and managing rights. Additionally, an interference management as well as the possibility of defining temporal usage rights and reclaiming them is suggested. Allowing flexible trading mechanisms lets the market decide about the success of the approach to transferring usage rights as introduced above. At the same time liberalization of spectrum is required to abolish restrictions on technologies and services associated with spectrum usage rights. Analysys Consulting (2004) recommends that this also be applied to existing spectrum licenses and suggests the reconfiguration as aggregation or partitioning of existing usage rights. A partitioning of usage rights is for example possible depending on time, frequency or geographical location.

3.1.2 Underlay and Overlay Spectrum Sharing

Open access to most of the radio spectrum, even spectrum licensed for a dedicated technology, is only permitted by radio regulation authorities for radio systems with minimal transmission powers in a so-called underlay sharing approach, as illustrated in Figure 3.1 (together with overlay sharing – for an illustration of overlay sharing see also Figure 3.3). The simultaneous uncoordinated usage of spectrum in the time and frequency domain uses techniques to spread the emitted signal over a large band of spectrum, so that the undesired signal power seen by the incumbent licensed radio devices is below a designated threshold. Spread spectrum, Multi-Band Orthogonal Frequency Division Multiplex (OFDM) or Ultra Wideband are examples of initial approaches of such techniques. To reduce potential interference, the transmission power can be strictly limited in underlay spectrum sharing.

In 2003 the Spectrum Policy Task Group of the FCC suggested (FCC, 2003b) the interference temperature concept for underlay spectrum sharing to allow low-power transmissions in licensed (used) bands. The FCC suggested allowing secondary usage of shared spectrum

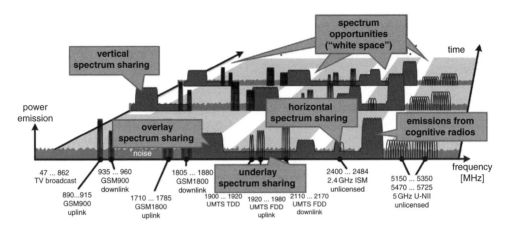

Figure 3.1 Opportunistic (secondary) spectrum usage with underlay spectrum sharing (low power) and overlay spectrum sharing (higher power, where spectrum opportunities, i.e. white space exists). Such methods for spectrum sharing could be combined and exploited by a cognitive radio in the future (Mangold *et al.*, 2005b)

if the interference caused by a device is below a sufficient threshold at the not to be interfered, primary receiver. The FCC claims that often, between the original noise floor and the licensed signal of the incumbent radios, there lies an interval that could be exploited, which was identified as 'new opportunities for spectrum use' (FCC, 2002, 2003b) and which is illustrated in Figure 3.2. This space refers to the power level of the signals at the receiver in a specific band at a geographical location.

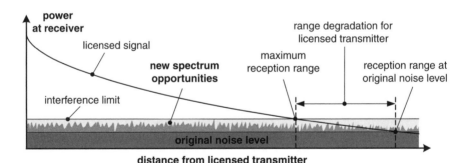

Figure 3.2 Underlay spectrum sharing corresponding to the interference temperature concept of the FCC (2003b). It allows secondary usage of licensed (used) spectrum with low power transmissions and refers to a well-defined space between original noise floor and licensed signal of the incumbent radios

Only a very small fraction of the radio spectrum is openly available as frequency band for unlicensed operation. Nevertheless, having these bands enables immense economic success of wireless technologies like the popular WLAN IEEE 802.11. On the other hand, the actual availability of new spectrum is a seemingly intractable problem. Cognitive radios use flexible spectrum access techniques for identifying under-utilized spectrum and avoiding harmful interference to other radios using the same spectrum. Such opportunistic spectrum access to under-utilized spectrum, whether or not the frequency is assigned to licensed, primary services, is referred to as overlay spectrum sharing.

Overlay sharing requires new protocols and algorithms for spectrum sharing. Additionally, spectrum regulation is impacted upon, especially in case of vertical spectrum sharing, which is discussed later in this chapter. The operation of licensed radio systems should not be interfered with when identifying spectrum opportunities and during operation in licensed spectrum. Dynamic frequency selection is a simple example of how unlicensed spectrum users (IEEE, 1999) share spectrum with incumbent licensed users (radar stations) using overlay sharing.

3.1.2.1 Ultrawideband as Underlay Sharing Approach

Ultrawideband (UWB) enables underlay spectrum sharing. In short, UWB is a transmission technique using pulses with a very short time duration across a very large frequency band of spectrum. An alternative and commercially attractive realization is based on multiband-OFDM (Hiertz et al., 2005a). In contrast to many other radio frequency communication techniques, UWB does not use RF carriers. Instead UWB uses modulated high frequency pulses of low power with a duration of less than 1 nanosecond. From the perspective of other communication systems, the UWB transmissions are part of the low power background noise. Therefore UWB promises to enable the usage of licensed spectrum without harmful interference to primary communication systems. UWB is often discussed in the context of the interference temperature concept of the FCC introduced above.

UWB coexistence and UWB-based cognitive radios are discussed by Lansford (Lansford, 2004). General methods of UWB are explored for using time, frequency, power, space and coding for eliminating interference to other devices, or to achieve graceful deterministic degradation. The application of UWB is limited to short range communication when broadband services should be realized, because of the strictly restricted average transmission power. The interference of UWB radios to incumbent radio systems is an ongoing discussion. For example, Pfletschinger and coworkers (Pfletschinger et al., 2005) claim that problems will arise due to co-channel interference between UWB and any future beyond-3G devices in indoor scenarios, in the case of the current FCC mask for UWB emission remaining as it is.

3.1.2.2 Opportunistic Spectrum Usage as Overlay Sharing Approach

Cognitive radios are radio systems that autonomously coordinate the usage of spectrum. They utilize radio spectrum when it is not being used by incumbent (primary) radio systems. Such unused radio spectrum is called 'spectrum opportunity'. The terms 'white spectrum/space' and 'spectrum hole' can be used equivalently. Figure 3.3(a) illustrates such white space in an artificial scenario. The figure indicates idle and busy (i.e., used) frequency spectrum over time/frequency. To use spectrum opportunities with overlay sharing, cognitive radios morph their transmission schemes such that they fit into the identified spectrum usage patterns, as illustrated in Figure 3.3(b). The figure indicates how a multicarrier wideband system such as that proposed for Ultra Wideband can change transmission powers, subcarrier spaces, and

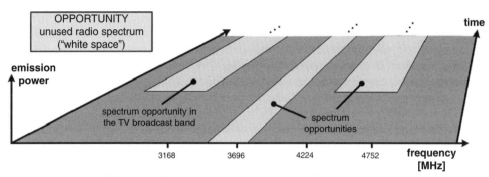

(a) White space (unused spectrum opportunities) in radio spectrum

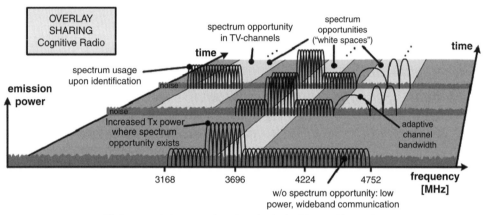

(b) Opportunistic usage (overlay sharing) with cognitive radio

Figure 3.3 Spectrum opportunities used in overlay spectrum sharing. To use spectrum opportunities with overlay sharing, cognitive radios morph their transmission schemes such that they fit into the identified spectrum usage patterns. The figure indicates how a multicarrier wideband system, such as that proposed for ultrawideband, can change transmission powers, subcarrier spaces, and subcarrier bandwidth to optimize spectrum usage. A spectrum opportunity may be defined by frequency band, the time interval when the spectrum is available, and geographic location. It is a radio resource that is either not used by licensed radio devices, or used with predictable patterns such that idle intervals can be detected

subcarrier bandwidth to optimize spectrum usage. For this, spectrum opportunities have to be identified that have high reliability, and their usage needs to be managed in a distributed and coordinated way. A spectrum opportunity is defined by frequency, time, and location. It is a radio resource that is either not used by licensed radio devices, or is used with predictable patterns, such that idle intervals can be detected.

The predictability and dynamic nature of spectrum usage contribute to the challenge of identifying spectrum opportunities accurately: the less frequent and more predictable the spectrum usage by primary radio devices, the higher the success in identification and efficiency of opportunistic usage by cognitive radios.

The predictable spectrum usage can be regarded as contribution to cooperation. This will be explained in Section 3.1.4.

Characteristic patterns in spectrum usage and characteristic signal features of the radio signals transmitted by primary radio systems facilitate an improvement of spectrum opportunity identification.

Different spectrum usage patterns and their classification as spectrum opportunities are shown in Figure 3.4. In this figure, time is progressing from bottom to top, frequency increases from right to left. Here we see patterns of random IEEE 802.11a/e spectrum usage (in three channels of the unlicensed 5-GHz band and the frequencies above), in parallel with a predictable, deterministic spectrum usage.

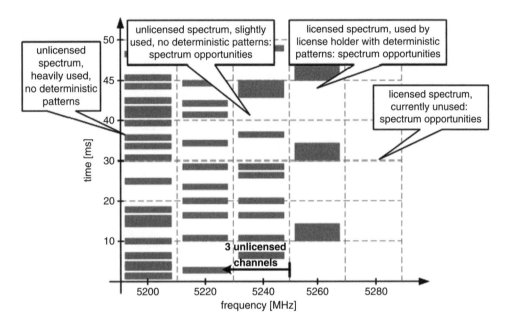

Figure 3.4 Spectrum usage example at 5 GHz. Three 802.11a/e channels and the frequencies above are depicted. A cognitive radio identifies spectrum opportunities. The graph was created with the help of event-driven simulation. Note that the illustrated spectrum usage at 5.26 GHz does not reflect a real-life application

3.1.2.3 Towards Our Vision of True Cognitive Radio

The overlay spectrum sharing with licensed radio systems requires not just fundamental changes in spectrum regulation; additionally, new algorithms for decision making and mutual coordination when sharing spectrum are necessary, which must reflect the different regulatory priorities for spectrum usage of the licensed, i.e. incumbent, and the unlicensed radio systems. To reflect this priority, the terms primary and secondary radio systems are used for the licensed and unlicensed radio systems respectively.

We are slowly entering the discussion of what true cognitive radio will look like. Overlay spectrum sharing must somehow be coordinated, otherwise regulators would not permit such ambitious new approaches. Decisions as to when to vacate, of what a spectrum opportunity may be, as well as mutual coordination and information exchanges between devices or dissimilar networks, may be centrally coordinated or may be made by the devices themselves. If radio devices make such decisions, regulators want to trace how they came to their conclusions – and this is exactly what cognitive radio will have to provide. See Chapter 7 for a detailed discussion of this vision.

3.1.3 Vertical and Horizontal Spectrum Sharing

Cognitive radios will have to share spectrum either (i) with unlicensed radio systems, or (ii) with licensed radio systems that are typically designed for exclusive use of otherwise unused spectrum.

The sharing of licensed spectrum with primary radio systems is referred to as vertical sharing, as indicated in Figure 3.5, and the sharing between radio systems with a similar regulatory priority (as for example in unlicensed bands) can be referred to as horizontal sharing. We have used these terms already in previous chapters. These terms of horizontal and vertical spectrum sharing are first mentioned by Kruys (Kruys, 2003). Another example horizontal spectrum sharing is the usage of the same spectrum by dissimilar cognitive radios (operated, for example, by different or even competing network operators) that are not designed to communicate with each other. These dissimilar cognitive radio systems have the same regulatory status, i.e. similar rights to access the spectrum, comparable to the coexistence of devices operating in unlicensed spectrum.

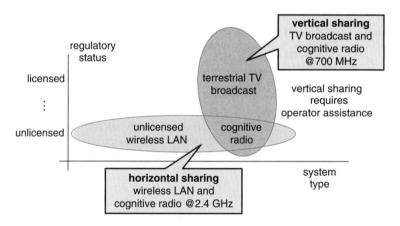

Figure 3.5 Cognitive radios share spectrum with different radio systems. Depending on the regulatory status, this is referred to as vertical or horizontal spectrum sharing

Vertical and horizontal sharing requires the capability to identify spectrum opportunities as introduced above. Cognitive radios are able to operate without harmful interference in sporadically used licensed spectrum while requiring no modifications in the primary radio system. Nevertheless, in order to protect their transmissions, licensed radio systems may assist cognitive radios to identify spectrum opportunities in vertical sharing scenarios. We refer to this support by a network operator as 'operator assistance'.

In horizontal sharing, the cognitive radios autonomously identify opportunities and coordinate their usage with other cognitive radios, usually in a distributed way. To avoid chaotic and unpredictable spectrum usage as in today's unlicensed bands, advanced approaches such as spectrum etiquette are helpful – this will be described in Chapter 7.

3.1.4 Coexistence, Coordination and Cooperation

In literature about spectrum sharing, the terms coexistence, coordination, and cooperation are often used differently. The three terms are important in the context of QoS support. A definition is given below.

Means for coexistence in spectrum sharing usually target interference mitigation and avoidance. In a distributed environment with the absence of a centrally coordinating entity, exchange of information through communication between competing devices is required anyway. This can be realized even between dissimilar radio systems by creating predictable radio emission patterns – patterns that signal levels of aggressiveness. This works as long as the resource demands of the individual radio systems do not exceed the available resources. Under severe competition for accessing the shared spectrum, no QoS support is possible due to the missing coordination among the coexisting, often dissimilar, radio systems like Wi-Fi and Bluetooth. Today's implemented approaches for coexistence offer limited interference prevention. Coexistence can be achieved with the help of spectrum etiquette, or alternatively by implementing new communication protocols and defining common spectrum coordination channels as introduced below.

Mutual coordination, either centralized or decentralized, is required in spectrum sharing to enable the support of QoS. QoS support refers in this context to the exclusive usage of spectrum for a predictable time and a predetermined duration. The intentional usage of deterministic patterns when allocating spectrum can be regarded as cooperation. These deterministic patterns help to increase the accuracy for competing radios for identification of spectrum opportunities and enable distributed coordination.

Cooperation is helpful for self-configuring networks of mutually coordinated radios in a distributed communication environment. Under cooperation, devices delimit their spectrum usage and carry each other's traffic in hope of gaining from potential cooperation when all radios participate. Cooperation comes along with the risk of being exploited by defecting, selfish, myopic radios. This usually results in disadvantage for the cooperating radios.

In distributed environments, cooperation can be created and enforced through protocols, either as part of a standard, or realized as spectrum sharing etiquette.

3.2 Horizontal Spectrum Sharing

In horizontal sharing, devices identify autonomously the spectrum opportunities and coordinate their usage in a decentralized way. To avoid chaotic and unpredictable spectrum usage as in today's unlicensed bands, the advanced approaches discussed in the following are helpful.

3.2.1 Coexistence

The amount of flexibility allowed in designing access protocols is imposed by regulating authorities for the possibility of radios to coexist, share spectrum or even interoperate for coordination. Peha (Peha, 2000) has therefore defined 'rules of coexistence' considering the level of flexibility from the least to the most restrictive:

Spectrum usage is not constrained. This approach has the advantage of stimulating innovations but has the big disadvantage of being only applicable in scenarios where less spectrum utilization can be expected. No QoS guarantee can be given and efficient spectrum usage is not encouraged.

Constraining spectrum usage is limited to the transmission power. A maximum power level for radio emissions is defined. Finding the adequate level may be difficult and depends on the application scenarios of the envisaged radios operating in these frequencies. Limiting transmission power increases the reuse of spectrum but more radios are required to cover a given area.

Constraints on access protocols are imposed, which require no communication between coexisting devices. Accurately defined etiquettes for spectrum usage can facilitate the support of QoS, increase spectrum efficiency, and add fairness to spectrum sharing. These etiquettes enable distributed coordination and require cooperation as introduced above.

A minimum standard is required for operation. A common signaling channel used by all radios operating in a shared frequency band is an example of this. The common signaling channel enables mutual coordination and thus increases the level of QoS.

The interoperation of dissimilar radio networks is part of their standard. The Central Controller Hybrid Coordinator (CCHC) concept for interoperation of IEEE 802.11 and HiperLAN/2 (Mangold et al., 2001b) or the Base Station Hybrid Coordinator (BSHC) concept for integrating IEEE 802.11(e) in the frame structure of IEEE 802.16 (WiMAX) as outlined in Berlemann et al. (Berlemann et al., 2006b) are examples of this.

In the following section several approaches to coexistence are introduced that do not require communication.

The FCC demands (FCC, 2003a) that devices operating in the U-NII band use Dynamic Frequency Selection (DFS) to protect radar systems from interference. DFS lets the transmitter dynamically switch to another channel whenever a certain condition is met. This condition is, for example, a threshold level like $-62\,\text{dBm}$ for devices with a maximum EIRP of less than 200 mW. Once a radio signal is detected, the channel must not be utilized for a certain period of time. Further, before initiating transmission a DFS using device senses the available spectrum for unused spectrum and accesses only these channels. The FCC requires that the DFS aims at a uniform spreading of load over the available channels. Additionally, a continuous monitoring of spectrum during operation is demanded. The possibilities and limitation of DFS for cognitive radios and policies are discussed by Horne (Horne, 2003). For fundamental details on DFS and radio networks using DFS see Walke (2002).

Transmit Power Control (TPC) is a mechanism that adapts the transmission power of a radio to certain conditions, e.g. a command signal from a communication target when the received signal strength falls below a predefined threshold. Corresponding to FCC (FCC, 2003a) in the U-NII band, a radio reduces its transmission power by 6 dB when TPC is triggered. Only low-power radios operating at power levels higher than 500 mW require TPC in the U-NII band.

3.2.2 Centralized Spectrum Coordination for Horizontal Sharing

One widely proposed approach to spectrum sharing is the usage of a Common Spectrum Coordination Channel (CSCC). The basic idea of a CSCC is to standardize a simple common protocol for periodically signaling radio and service parameters, regarded as spectrum etiquette mechanism (Raychaudhuri and Jing, 2003). The CSCC enables coordination, through mutual observability, between different neighboring radio devices via a simple common protocol. As shown by Raychaudhuri and Jing (2003) at the contention example of 802.11b and Bluetooth devices the CSCC approach impacts the complete protocol stack (e.g. physical layer, MAC layer, packet formats).

In general, a common control channel as part of the shared spectrum is highly vulnerable to interference so that the complete network might be disrupted through deliberate jamming or coexisting radios operating in the same spectrum.

A permanently available signaling channel shared by several networks is also suggested (Hunold et al., 2000). A Network Access and Connectivity Channel (NACCH) is introduced to enable communication between different networks. In this way, a larger network is formed, where users have universal access and roaming support. This approach extends the exchange of control information between networks for coordinating spectrum usage with the aspect of transferring user data between networks over a common radio channel.

3.2.2.1 Brokerage-based Horizontal Spectrum Sharing

Spectrum can be regarded as economic goods that are traded by a broker. Auctions are an efficient way to determine the value of spectrum among many parties. An action can be regarded as a partial information game in which the real valuation a bidder gives to spectrum is hidden to the broker and the other bidders. Auction theory is a very well developed field of research in economics as well as in wireless communication. Courcoubetis and Webber (2003) give a technology-independent introduction and some theoretical results of spectrum auctioning.

The different treatment of spectrum in the context of brokerage-based spectrum sharing leads to dissimilar approaches. The auctioning mechanisms introduced by Maheswaran and Basar (2003) consider spectrum as a public resource. Price and demand functions characterize the optimal response functions of the bidders leading to a unique Nash equilibrium for an arbitrary number of agents with heterogeneous quasi-linear utilities, contrary to Grandblaise et al. (2005) where another business model is applied. They discuss spectrum sharing between different operators providing their spectrum to a common spectrum pool. An operator temporarily leases spectrum from a broker who controls this pool of spectrum and performs the auctions. An analytical investigation of the problem and solution approach is described by Rodriguez et al. (2005): an operator of a Code Division Multiple Access (CDMA) cell populated by delay-tolerant terminals operating at various data rates, tries to optimize revenue, given an amount of spectrum depending on its own pricing policies. Brokerage-based spectrum sharing implies substantial periodical signaling, depending on the number of participating parties. The automatic bidding through agents located in the MAC layer is also discussed by Kloeck et al. (2005) for an OFDMA/TDD system like IEEE 802.16. Piggybacked multi-unit, sealed-bid, actions are suggested in order to take heavy time and signaling constraints into account.

It can be questionable, in general, whether or not a QoS guarantee can be given when using spectrum based on auctions. Therefore, a customer-oriented network operator might not be willing to provide its spectrum to the pool even if in the short term the selling of its capacity leads to higher revenues.

3.2.2.2 Inter-operator Horizontal Spectrum Sharing

Another approach to DSA is introduced by Pereirasamy *et al.* (2005) as Dynamic Inter Operator Spectrum Sharing. There, based on spectrum shared in UMTS Terrestrial Radio Access, Frequency Division Duplex (UTRA FDD), each operator deploys its own independent radio access network. This work was initiated by Pereirasamy and coworkers (Pereirasamy *et al.*, 2004) where a shared UMTS FDD network is assumed. From the regulatory perspective, inter-operator spectrum sharing is limited to one network and a frequency band licensed to be used by this single network is dynamically divided between several operators. The consideration of a synchronous inter-operator system requires heavy cooperation, which is very unlikely for competing operators. Further, it has been shown (Pereirasamy *et al.*, 2005) that only partial spectrum sharing leads to favorable capacity gains, and these depend heavily on network parameters such as cell radius, and transmission power.

3.2.3 Spectrum Sharing Games

The application of solution concepts derived from game theory for the analysis and realization of spectrum sharing has many facets. The corresponding game models and their characteristics differ significantly with application scenario. The horizontal spectrum sharing of equally righted cognitive radios is suitable for modeling as a game of interacting players.

Potential game models in which networks of cognitive radios may alter transmitted energy and signature waveform have been analyzed by Neel and coworkers (Neel *et al.*, 2002a,b). Specific conditions are delineated for which the models apply. These models are used to identify the steady-state conditions of these networks in the presence of cognitive radios that are making myopic decisions as an initial step in network planning. Different convergence dynamics in cognitive radio networks have been examined (Neel *et al.*, 2004).

Several different distributed power control algorithms are examined in game models of different complexity levels. The convergence process and the steady-state behavior of cognitive radio networks are analyzed in modifying the objective functions to prevent the application of adaptation algorithms that result into suboptimal game outcomes.

The selfish behavior of nodes in CSMA/CA networks has been investigated (Cagalj *et al.*, 2005). In CSMA/CA the protocols rely on random deferment of packet transmission. By applying a dynamic games model, the conditions for stable and optimal operation in the presence of selfish nodes are analyzed.

The channel assignment in a Wireless LAN has been modeled as a game (Halldorsson *et al.*, 2004) in which the players are service providers or access points. There, stable points of operation are identified with the solutions to a maximal coloring problem in an appropriate graph. A 'price of anarchy' (Halldorsson *et al.*, 2004) is introduced as value for the confession to the distributed channel assignment in comparison with a centralized one. In the case of easily realizable bargaining procedures, this price of anarchy is bounded to a constant.

A game theoretic formulation of the adaptive channel allocation problem for cognitive radios resulting into rules for spectrum etiquette is discussed (Nie and Comaniciu, 2005). Cognitive radios measure the local interference temperature on different frequencies and may adapt their transmission rate to channel quality (in using adaptive channel coding) or switch to a different frequency channel. It was shown that the corresponding game formulation converges to a deterministic channel allocation scheme. Additionally, adaptive protocols and a learning algorithm were designed and their convergence behavior and tradeoffs discussed.

The competition between independent radio systems for allocating a common shared radio channel can be modeled as a stage-based game: players, representing radio systems, interact repeatedly in radio resource-sharing games without direct coordination or information exchange. Solution concepts derived from game theory allow the analysis of such models under the microeconomic aspects of welfare.

3.3 Vertical Spectrum Sharing

All approaches to horizontal spectrum sharing can be used for vertical spectrum sharing when being combined with an accurate identification of spectrum opportunities. When multiple radios simultaneously regard the same spectrum as unused by the incumbent radio system, the access by the secondary radios needs to be coordinated (centralized or distributed) to enable the support of QoS on the one hand and to increase efficiency of spectrum usage on the other.

Vertical spectrum sharing can be realized in different ways. Beacon signal or busy tone at a dedicated frequency for signaling permission and/or prohibition of secondary operation in licensed spectrum is one approach to vertical spectrum sharing. A licensee may sell temporarily under-utilized spectrum for secondary usage to increase its revenue. Mechanisms for this have been introduced previously in Section 3.1.1. More complex approaches to vertical sharing like a common control channel or the policy-based secondary spectrum usage on the basis of spectrum observation are discussed below.

3.3.1 Reuse of TV Bands for Vertical Spectrum Sharing

The technology for terrestrial TV broadcasts is currently digitized. This process will be finalized in the near future (for example in the US in 2009; in Germany latest by 2010). This digitalization improves the utilization of spectrum, resulting in a reduction of required spectrum while the number and quality of the TV channels remain unchanged. The usage of the corresponding frequency band is reorganized at the same time in many regulatory domains worldwide. As every broadcast site has to serve a large coverage area, radio transmission is at high power to guarantee reliable reception throughout the complete coverage area. For many receivers this implies a robustness to interference in case of proximity to the broadcast site, as the signal is received at a higher power than required. Thus, reliable operation is possible even if cognitive radios emit some level of interference. Additionally, terrestrial TV broadcast sites never change their location, and frequency allocations to broadcast channels remain virtually constant over a period of years, which simplifies the identification of spectrum opportunities.

It is therefore envisaged that such unlicensed reuse of the entire TV broadcast band for cognitive radios that scan all TV channels throughout the band and operate only upon identification of spectrum opportunities (FCC, 2004c) be allowed. The working group 802.22 of the IEEE takes this idea and is working towards the standardization of the unlicensed secondary access to TV bands, as outlined in Section 4.5.3.

This scenario is illustrated by Figure 3.6, which shows two adjacent TV broadcast sites and two independent pairs of communication cognitive radio devices. The cognitive radios identify locally under-utilized spectrum, in this case unused TV channels, as spectrum opportunities. After some knowledge dissemination and negotiation, the pairs of cognitive radios communicate using these opportunities, while frequently scanning the spectrum for signals from primary radio systems.

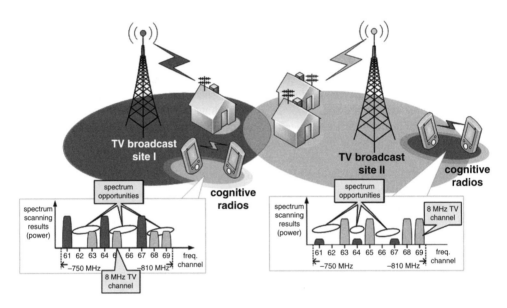

Figure 3.6 Cognitive radios operating in frequency bands of TV and radio broadcasts. At different locations the cognitive radios identify different frequencies as unused and regard them as spectrum opportunities

There are other approaches for improving the spectrum usage efficiency of the TV broadcast spectrum. The World Radio Conference 2007 (WRC07) has discussed opening spectrum below 1 GHz to new cellular networks, motivated by the ongoing rearrangement of the terrestrial TV spectrum at 470–862 MHz from analog to digital TV. With digital TV, around 100 MHz will be available in segments of 6 MHz or 8 MHz channels previously occupied by analog terrestrial TV. This spectrum allows cost-effective solutions for networks with nationwide coverage, including areas with low population density, because of the favorable propagation characteristics of lower frequencies. The spectrum is of particular interest for operators that are obliged to provide nationwide wireless broadband services.

3.3.2 Spectrum Pooling and a Common Control for Vertical Spectrum Sharing

Weiss and Jondral (2004), have developed an OFDM-based approach to secondary usage in overlay spectrum sharing, referred to as spectrum pooling. In a WLAN-like scenario, the spectrum measurements of mobile devices are gathered centrally by an access point. The unused spectrum of different owners is merged into a common pool that will be optimized for a given application. A licensed system hosts this common spectrum pool and users can temporarily rent spectrum during the idle periods of the licensed users. Although Weiss and Jondral introduce a speeding-up protocol to bypass the MAC layer and just use the physical layer for signaling, an essential weakness of spectrum pooling is only mitigated. The central gathering of measurement information takes considerable time and management effort.

In the context of vertical spectrum sharing, Cabric *et al.* (2005) introduced opportunistic spectrum usage creating a virtual unlicensed band, also referred to as a spectrum pool, with the help of a common control channel. A hierarchical control channel structuring is suggested for differentiating between a universal control channel, used by all groups for coordination, and separate group control channels used by members of a group.

A dedicated control channel located in licensed spectrum is suggested by the DARPA XG Program (DARPA, 2003, 2004b) to enable coordination in shared spectrum. This simple approach has several disadvantages, as a fixed licensed channel is required. The licensing of such a channel is a long-winded and expensive process. The licensed channel has a fixed bandwidth and is therefore limited in scalability.

3.3.3 Operator-assistance in Vertical Spectrum Sharing

The rationale of cognitive radio discussed above assumes a decentralized organization of spectrum access. A multi-radio device could, for example, be managed by multiple in-parallel existing cognitive radio entities that may coordinate their usage of available spectrum and radio interfaces. A cognitive radio senses its radio environment and coordinates spectrum sharing with neighboring cognitive radios in a distributed way. Practically, such an operation is locally limited to short range communication in order to avoid interference to incumbent radio systems.

'Operator-assistance' (Mangold *et al.*, 2006b,c) introduces centralized means for coordinating spectrum access to the distributed paradigm of cognitive radio. Such a centralized approach has the advantage of facilitating the regulator's control of spectrum usage, and allowing them to direct how the spectrum is used. Such a control is especially important in vertical spectrum sharing. In contrast, the decentralized approach is more flexible, which is especially helpful in horizontal spectrum sharing. To protect licensed radio systems in vertical spectrum-sharing scenarios, but to gain from the flexibility of decentralized approaches, supplementary radio systems owned by operators may assist cognitive radios in identifying under-utilized spectrum. In this way, dynamic coordination of spectrum usage is facilitated with the help of a radio network that has clearly defined site locations and a trusted operator.

Signaling mechanisms based on denial and grand beacons broadcast by an incumbent operator enable the control and coordination of secondary radio systems operating in the operator's frequency band, without interfering with its primary radio systems. Such a beaconing concept as an exemplary approach to operator assistance is discussed in detail later in this book.

A multitude of new additional services can be offered by an operator in the context of cognitive radio, with the help of a common coordination channel (i.e. pilot channel) for broadcasting free frequency bands and technologies available for using these bands. This might also imply reconfiguration information for cognitive radios as essential part of a reconfigurable, cognitive radio network allowing a balancing of devices operating in locally available networks. Costs, traffic load and QoS constrains are example parameters that might be taken into account when cognitive radios choose a transmission technology with the help of the operator.

An operator may also assist in coordinating the distributed spectrum access of cognitive radios. Such coordination reduces mutual interference among cognitive radios in mitigating, for example, the hidden station problem. In this way, QoS support for spectrum sharing cognitive radios is improved.

A reasoning service provided by an operator may help to reduce the complexity of the decision logic of cognitive radios. Multiple parameters (such as spectrum policies, the local spectral environment or QoS constraints) are to be combined when cognitive radios decide about spectrum allocation. A cognitive radio may therefore delegate the decision taking to the operator's network infrastructure.

A trusted operator may further provide services for enabling individual spectrum trading, and authentication, bidding and accounting can be realized through an operator on behalf of a cognitive radio's owner to reduce further the device complexity.

3.3.4 Spectrum Load Smoothing for Vertical Spectrum Sharing

The application of waterfilling in the time domain enables a decentralized and coordinated, opportunistic usage of the spectrum. This is referred to as Spectrum Load Smoothing (SLS) (Berlemann and Walke, 2005; Berlemann et al., 2006c; Berlemann, 2006), and is discussed in detail in Section 6.3 of this book.

With SLS, competing radio systems aim simultaneously at an equal utilization of the spectrum. In observing the past usage of the radio resource, the radio systems interact and redistribute their allocations of the spectrum under consideration of their individual QoS requirements. Due to the principles of SLS, these allocations are redistributed to less utilized or unallocated spectrum. QoS requirements of the coexisting networks are considered. Further, SLS allows optimized usage of the available spectrum: an operation in radio spectrum that was originally licensed for other communications systems, is facilitated, as the SLS implicitly achieves usage of unused spectrum and its release in case it is needed again.

3.4 Taxonomy for Spectrum Sharing

We refine John M. Peha's appropriate 'taxonomy for spectrum sharing' (Peha, 2005) by extending it with respect to vertical and horizontal spectrum sharing. The different approaches to regulation of spectrum usage which have been introduced in this chapter are summarized in Table 3.1, in taking the aspect of QoS into account. Peha's understanding of a cooperative meshed network is matched thereby to the cognitive radio network vision as introduced above. The table separates regulation options into primary and secondary spectrum usage. A QoS guarantee always requires some degree of exclusiveness, but if a guarantee is not

Table 3.1 Regulation options for (a) primary and (b) secondary spectrum usage (Peha, 2005)

(a) Regulation options for primary spectrum usage

Regulator controls access	Licensee controls access	Application requirements
Traditional licensing	Spectrum manager makes guarantees	Guaranteed QoS
Unlicensed band, regulator sets etiquette	Spectrum manager sets etiquette, no QoS guarantee	No QoS support, coexistence, horizontal spectrum sharing
Cognitive radio, regulator sets protocol	Cognitive radio, licensee sets protocol	QoS support, cooperation, horizontal spectrum sharing

(b) Regulation options for secondary spectrum usage

Regulator controls access	Primary licensee controls access (secondary market)	Application requirements
Not possible	Licensee guarantees QoS	Guaranteed QoS
Unlicensed underlay with opportunistic access	Secondary market with overlay opportunistic access	No QoS support, coexistence, vertical spectrum sharing
Interruptible secondary operation, regulator sets cooperation protocol	Interruptible secondary operation, regulator sets cooperation protocol	Interruptible QoS support, cooperation, vertical spectrum sharing

required, primary systems may share spectrum. Coexistence is less adequate even to support QoS, while cooperation increases the level of possible QoS support. Regulation authorities can delegate the control of spectrum access to one or to multiple private entities in order to enable spectrum trading in the secondary market. A so-called spectrum manager inherits the role of the regulator in this context. Secondary usage might be allowed for underlay or overlay spectrum sharing, provided that secondary radio systems defer from spectrum utilization whenever the license-holding primary radios access their spectrum. Secondary radios can try to coexist with primary radios without interfering with them. Cooperation between secondary and primary radios enables the secondary radios to support QoS with known interruptions. Secondary radio systems are only able to guarantee QoS if the primary radio systems commit themselves not to interfere. This commitment of the licensee introduces trading of spectrum.

4

Towards Cognitive Radio – Research and Standardization

Commercially applied wireless communication systems are based on perfectly understood technologies with standardized transmission schemes and protocols. To guarantee a full inter-operability of their radio systems, vendors and chipmakers need to agree on a minimum set of technology as part of a communication standard. Operators will only invest in a technology that is established and provided by more than one vendor in order to retain their independence from the distributors of the used technology. Hence, operators are also interested in communications standards. Hence, it was clear from the beginning that cognitive radio should focus on standardization, and so many research activities and many industry alliances focus on extensions of existing standards, such as the IEEE 802 family, towards full cognitive radio and towards dynamic spectrum access for more flexibility in spectrum access. Cognitive radio will therefore not replace established or commercially successful systems. Instead, it will provide them with incremental improvements in the needed flexibility. We call such improvements 'enablers' in the following chapters. Before we discuss such enablers, current major activities in the realization of cognitive radio (standardization, research programs, and conferences) are listed below. Some standards, namely IEEE 802.11n for Wi-Fi MIMO and channel bonding, as well as IEEE 802.11s for Wi-Fi mesh networks, are discussed in more detail. Further, we discuss the WiMAX IEEE 802.16 approach, including IEEE 802.22, which describes a new standard for radio systems operating in the TV broadcast spectrum.

4.1 Research Programs and Projects

In the recent years, various research organizations have initiated programs and projects aimed at the improvement of flexibility in wireless communication. In the US, the SDRforum is targeting a reconfigurable system architecture for wireless networks and user terminals on the basis of Software Defined Radio (SDR). The IST-projects and E^{2}R of the Sixth Framework research funding Program (FP6) of the European Union, belonging to the Wireless World Research Initiative (WWI), both concentrate on different levels of flexible radio interfaces

Cognitive Radio and Dynamic Spectrum Access   L. Berlemann and S. Mangold
© 2009 John Wiley & Sons, Ltd

and network architectures. E^2R, just as DARPA XG, is working also on flexible and dynamic spectrum usage and related impacts on spectrum regulation.

4.1.1 DARPA Next Generation Communications Program, XG

The Defense Advanced Research Projects Agency (DARPA) NeXt Generation Communication (XG) Program, financed by the US government, aims at developing a *de facto* standard for cognitive radio, and dynamic spectrum regulation (DARPA, 2003, 2004b, 2008). The DARPA XG Program can be considered to be the number one source of innovation and the major driver of cognitive radio. The program schedule is divided into three phases: (i) technology investments (from 2002–03); (ii) system and protocol design (from 2003–05), and (iii) system development and demo (from 2005–08). The key principles are independence from specific radio architectures, interference prevention in dynamic spectrum access, and to separate policies from engineering. Thereby DARPA XG focuses on spectrum awareness, adaptive network operation, interference effects assessments and behavior.

This is reflected in the XG concept of cognitive radio, which is based on so-called 'abstract behaviors, protocols, and a policy language.' The reasons for this approach are mainly 'flexibility,' 'long-term impact,' and the need for traceability, i.e. regulatory approval of the rules being used by cognitive radios. In other words, behaviors are used instead of detailed descriptions of a standardized protocol, or a set of different standardized protocols, to allow regulators and industry to align dynamically future regulatory requirements and rules for spectrum usage with existing and emerging technologies for future radio systems. Policies use a policy meta-language as utility. There is a direct association between policies and technical constraints. Abstract behaviors are derived from policies and a behavior is composed of core behaviors. Protocols are derived from behaviors, realized by real implementation. DARPA has published its policy language and first draft implementations for traceable decision making of cognitive radios by BBN Technologies (DARPA, 2003, 2004a, 2004b). This work has been continued/replaced by SRI (Elenius *et al.*, 2007; SRI, 2008; Wilkins *et al.*, 2007) with a less comprehensive policy language framework that promises an easier prototypical implementation of reasoning.

4.1.2 National Science Foundation's Project GENI

The Global Environment for Network Innovations (GENI) is a US Research program and experimental facility initiated in 2006 (GENI, 2008). GENI is an open, large-scale, realistic experimental facility with a central goal, namely changing the nature of networked and distributed systems in moving from a fully empirical to a rigorously understood design process.

GENI is concentrating on the impact of cognitive radio on the architecture, design, and implementations of networks in general, summarized under the umbrella of Cognitive Radio Networks. Evans *et al.* (2006) show several examples of how techniques and mechanisms necessary to cognitive radio networks will have an influence on GENI. An experimental infrastructure consisting of a flexible cognitive radio network test environment, an open air cognitive radio network test environment, a software defined radio experimental platforms, plus simulation and emulation tools, will be implemented by GENI project partners.

4.1.3 European Project E³

The End-to-End Efficiency (E³) project is a large project funded by the EC in the Seventh Framework Program (E3, 2008). E³ is aiming at integrating cognitive wireless systems into Beyond 3G (B3G) technologies, evolving current heterogeneous wireless system infrastructures into an integrated, scalable and efficiently managed B3G cognitive system framework. The key objective of the E³ project is to design, develop, prototype and showcase solutions in order to guarantee interoperability, flexibility and scalability between existing legacy and future wireless systems, to manage the overall system complexity, and ensure convergence across access technologies, business domains, regulatory domains and geographical regions. E³ is a two-year project initiated in January 2008 with 21 partners from industry and academia. It can be regarded as the successor of the Sixth Framework Program IP E²R (Bourse *et al.*, 2004; 2005). The technical focus of E³ is concentrating on a cognitive radio architecture (CRA), a cognitive pilot channel (CPC), dynamic spectrum management, collaborative cognitive radio resource management, self-configuration and self-optimization, spectrum sensing, autonomous cognitive radio functionalities, a domain specific language (DSL), and the E³ prototyping system.

4.1.4 European Project WINNER+

The Wireless World New Radio Initiative+ (WINNER+), as successor to WINNER and WINNER II (Mohr, 2005), is a two-year European research project that was initiated in Spring 2008. WINNER+ is a 'Cooperation for a sustained European Leadership in Telecommunications' (CELTIC) project joining forces from the infrastructure industry, telecoms, small/medium sized enterprises, universities, and research institutes. WINNER+ is targeting development, optimization and evaluation of a competitive IMT-Advanced candidate proposal by integrating innovative and cost-effective additional concepts and functions, and providing an evolutionary path towards further improved performance of IMT-Advanced. Spectrum sharing, reuse and flexible spectrum usage are some of the key innovations in the focus of WINNER+. Principles of cognitive radio are integrated into the centrally organized world of mobile cellular radio networks.

4.1.5 European Project WIP

The All-Wireless Mobile Network Architecture (WIP) project is an EU funded, Specific project of the Sixth Framework Program, initiated in 2006 (WIP, 2008). The objective of the WIP project is to design, implement, and experimentally validate an all-wireless interconnection architecture based on advanced wireless transmission techniques, mesh networking, cross-layer optimization, and mechanisms for seamless mobility. Such a wireless network must be self organizing in order to maximize the network capacity and capabilities. The project will develop a suitable measurement methodology and set up experimental prototypes to validate the proposed concepts.

To fulfill the vision of what is also called the 'radio internet', WIP will adopt a disruptive approach with respect to the current Internet: to rethink the overall architecture with wireless networks as the central technology and develop a new interconnection architecture founded on the principle of all-wireless networks, with some isolated wired long-haul links. By defining new topology and routing schemes, designing simple interconnection rules, elaborating new performance objectives, providing seamless mobility, and designing self-configuring and self-healing mechanisms, a large number of wireless access networks can be interconnected dynamically to provide wide coverage with a large capacity and ubiquity. The technical focus of WIP lies in hierarchical wireless mesh architecture, operator assisted cognitive radio, cross-layer design and optimization, new addressing and routing schemes, communities implemented on several layers, and experimentation testbeds.

4.1.6 European Project SOCRATES

SOCRATES (Self-Optimisation and self-ConfiguRATion in wirelEss networkS) is a European research project funded by the EC under the Seventh Framework Program (SOCRATES, 2008). SOCRATES is aimed at the development of self-organization methods to enhance the operation of wireless access networks by integrating network planning, configuration and optimization into a single, mostly automated, process requiring minimal manual intervention. The objective is to effect substantial operational expenditure (OPEX) reductions by diminishing human involvement in network operational tasks, while optimizing network efficiency and service quality.

4.1.7 European Project ROCKET

The EU research project ROCKET (Reconfigurable OFDMA-based Cooperative NetworKs Enabled by Agile SpecTrum Use) is a STREP of the Seventh Framework Program (ROCKET, 2008). ROCKET is targeting the development of mobile radio networks with bit rates higher than 100 Mb/s (Mbps) and with peak throughputs higher than 1 Gb/s (Gbps). A special focus is the provisioning of a homogeneous, high rate, coverage within the served area. The requirements of IMT Advanced and IEEE 802.16m are addressed by devising methods for improved opportunistic spectrum usage, advanced multi-user cooperative transmission and ultra-efficient MAC design. The main pillars of ROCKET are operation in wider channels, opportunistic spectrum usage, scattered multiband operation, advanced cooperative transmission/reception techniques, base stations and relay stations coordination, ultra-efficient MAC layer, and adaptive schemes and resource allocation.

4.1.8 European Project ORACLE

The key objective of Opportunistic Radio Communications in Unlicensed Environments (ORACLE) is to research, develop, and validate concepts, mechanisms and architectures for cognitive radio networks and terminals, and to demonstrate the socio-economical advantages of opportunistic spectrum usage (ORACLE, 2008). The development of agile terminals and mechanisms to facilitate access networks (i.e. the access points, base stations and/or Node-Bs) that can sense 'white spaces' in used spectrum and adapt their transmission characteristics to use these 'white spaces' in the radio spectrum, will provide one tool to optimize spectrum utilization. The aim of ORACLE is to develop sensing techniques, define decision making processes, propose adaptable baseband architectures, design protocols and algorithms that identify, disseminate and exploit space dimension opportunities in opportunistic radio networks and build up a prototypical hardware platform demonstrating opportunistic terminal and network features.

4.2 IEEE Coordination, and the Coexistence Advisory Group IEEE 802.19

The Institute of Electrical and Electronics Engineers (IEEE) is in the forefront of cognitive radio standardization. The industrial relevance of cognitive radio in the US is the result of regulation authorities that, for many years, advocated the liberalization of spectrum regulation. This resulted in the necessary required industrial momentum for the development and commercial realization of cognitive radio in the US. Many IEEE standardization bodies are working on incremental steps towards the vision of cognitive radio as summarized in Table 4.1.

Table 4.1 Overview of standardization activities towards cognitive radio within the IEEE

IEEE Standardization Body	Cognitive Radio Feature/Building Block
SCC 41/P1900	Series of general standards on cognitive radio
802.11n	Phased coexistence operation for phased 20/40 MHz channel usage
802.11s	Common channel framework for multi channel operation
802.11k	Measurement, reporting, estimation and identification of characteristics of spectrum usage
802.11y	High power contention based medium access and a flexible spectrum management framework
802.16h	Coexistence among license-exempt systems based on IEEE 802.16
802.16.2	Coexistence in IEEE 802.16 systems operating in licensed bands
802.19	Coexistence technical advisory group
802.22	Wireless rural access network for the (re-) use of TV bands with a radio technology similar to IEEE 802.16/WiMAX

Many of the wireless communication systems standardized by the IEEE are operating in unlicensed spectrum. Within unlicensed frequency bands, radio systems coordinate the usage of radio resources autonomously while operating.

The IEEE 802.19 working group calls itself Coexistence Technical Advisory Group. 802.19, and is aims at the development and maintenance of policies defining the responsibilities of IEEE standardization efforts to consider coexistence with existing standards and other standards under development. If required, 802.19 evaluates the conformity of the developed standard to coexistence policies and offers documentation of the coexistence capabilities to the public.

A framework for defining coordination rules for radio resource management, referred to as spectrum etiquette, is discussed later in this book. Spectrum etiquette rules for radio systems with different channel bandwidths are defined. The rules are based on a set of actions like channel selection and listen-before-talk. By evaluating the rules with help of simulations, Mangold and Challapali (2003) provide an initial approach towards a spectrum etiquette proposal.

4.3 IEEE SCC41/P1900

The IEEE Standards Coordinating Committee (SCC) 41 has initiated a series of standards, the IEEE P1900 series on next generation radio and advanced spectrum management, to stimulate the research and development of cognitive radio. SCC 41 evolved in April 2007 from the IEEE P1900 Standards Committee that was established in 2005. The scope of SCC 41 includes improvement of spectrum usage, new techniques and methods of dynamic spectrum access, interference management, coordination of wireless technologies, network management and information sharing. The SCC 41 is currently divided into 5 WGs:

(i)   IEEE P1900.1 on Terminology and Concepts for Next Generation Radio Systems and Spectrum Management;
(ii) IEEE P1900.2 on Recommended Practice for Interference and Coexistence Analysis;
(iii) IEEE P1900.3 on Recommended Practice for Conformance Evaluation of SDR Software Modules;
(iv) IEEE P1900.4 on Architectural Building Blocks Enabling Network-Device Distributed Decision Making for Optimized Radio Resource Usage in Heterogeneous Wireless Access Networks;
(v)   IEEE P1900.5 on Policy Language and Policy Architectures for managing cognitive radio for Dynamic Spectrum Access Applications.

4.3.1 IEEE P1900.1

As many research groups have defined the cognitive radio and related terms differently, P1900.1 is creating a glossary of terms and concepts in the context of cognitive radio. Technically precise definitions are standardized and explained in order to provide a common understanding of cognitive radio and to found a framework for the standardization efforts of the other IEEE SCC41 WGs. An example of this is the notation of vertical and horizontal spectrum that was contributed by the P1900.1 working group.

4.3.2 IEEE P1900.2

One major objective of cognitive radio is to improve the overall efficiency of spectrum usage. For this reason, the accurate measurement of interference has become a crucial requirement in the deployment and evaluation of cognitive radio technologies. P1900.2 is, therefore, engaged in developing interference analysis criteria. Additionally, a framework for measuring and analyzing the spectral coexistence between different radio systems is established. The resulting common standard platform shall facilitate the resolution of (inevitable and necessary) disputes on the introduction of cognitive radio technologies. The flexibility and adaptability of spectrum usage stimulate fear and doubts in cognitive radio, especially among incumbent license holders and conservative regulation bodies. The tradeoff between (i) interference prevention in adaptive spectrum usage, and (ii) cost through precise electrical components must be quantified. For this reason, uncertainty levels in measurements and thresholds of harmful interference are defined by P1900.2.

4.3.3 IEEE P1900.3

SDR is an important part of cognitive radio. The P1900.3 working group is developing test methods for conformance evaluation of software for SDR devices. A set of recommendations is defined to assure the coexistence and compliance of the software modules of cognitive radio devices before proceeding toward validation and certification of final devices. P1900.3 wants to increase confidence in SDR devices. SDR will comprise different layers of software with different functionalities, each to be validated on their conformance to regulatory and operational requirements. The P1900.3 working group is designing testing procedures that will comply with the semiformal software specifications in defining, for example, checkpoints and assertions that reflect the specification. Guidelines for operators and manufacturers of SDR technologies shall be established to enable licensing by regulation authorities.

4.3.4 IEEE P1900.4

The IEEE P1900.4 working group is standardizing means for distributed decision making in heterogeneous wireless networks in order to optimize the usage of radio resources. Therefore, network and device resource managers, as well as the communication exchange between them, are developed. In a first stage, the standard is limited to architectural and functional definitions while the corresponding protocols are defined in a later stage. The focus on cognitive radio capabilities in the context of heterogeneity in wireless access technologies differentiate this working group from other WGs of SCC41. The Cognitive Pilot Channel (CPC) concept is an example for the work of P1900.4. The CPC assists a reconfigurable and cognitive terminal to select a specific radio access technology in a wireless communication environment of different access networks with varying spectrum allocations.

4.3.5 IEEE P1900.5

A policy language (or a set of policy languages or dialects), together with a corresponding policy architecture, is being developed by the youngest member of SCC41, the IEEE P1900.5 working group. A policy framework is specified for an interoperable, vendor-independent, control of cognitive radio functionality and behavior for dynamic spectrum access. Initial efforts will concentrate on the features required for bounding a policy language to one or more policy architectures. Thereafter, concrete details with a special focus on interoperability are intended to be standardized. The policy-defined spectrum sharing and medium access for cognitive radios discussed later in .this book are based on similar principles and are closely related to P1900.5.

4.4 Wi-Fi Wireless Local Area Networks IEEE 802.11

4.4.1 IEEE 802.11k for Radio Resource Measurements*

IEEE 802.11k (IEEE, 2005c) is an amendment to the IEEE 802.11 base standard for radio resource measurements and was approved in May 2008. Various types of measurement are defined that enable wireless LAN stations to request measurements from other stations, for example, in order to measure how occupied a frequency channel is. The measurement results are reported back to the requesting station in standardized frames. It provides means for measurement, reporting, estimation and identification of characteristics of spectrum usage. 802.11k improves spectrum opportunity identification in unlicensed bands in unpredictable environments and is able to characterize the interference on different frequency channels.

The objective of 802.11k is to provide measurements and frame formats by which a radio station can initiate, measure, and assess the radio environment. Note that 802.11k is referring to radio resource measurements, and not radio resource management. Therefore, actions that make use of the new information are not defined, only the part of the entire management process defined by 802.11k that involves measurement, including requesting and reporting. To fulfill its objective, the 802.11k amendment defines different types of measurement, of which some are briefly described in the following.

802.11k enables a radio network to collect information about other access points (via beacon report), and about the link quality to neighbor stations (via frame report, hidden node report and station statistic report). 802.11k also provides methods to measure interference levels (via noise histogram report) and medium load statistics (via channel load report and medium sensing time histogram report).

With the beacon report, a measuring station reports the beacons or probe responses it receives during the measurement duration. With the frame report, a measuring station reports information about all the frames it receives from other stations during the measurement duration. With the channel load report, a measuring station reports the fractional duration over which the carrier sensing process, i.e., Clear Channel Assessment (CCA), indicates that the medium is busy during the measurement duration. In the noise histogram report, a measuring station reports non-802.11 energy by sampling the medium only when CCA indicates that no 802.11 signal is present. With the hidden node report, a measuring station reports the existence, and frame statistics of hidden nodes detected during the measurement duration. In the station statistical report, a measuring station reports its statistics related to link quality and network performance during the measurement duration.

The optimal measurement durations, the reliability of results, and the relevance of the results for the time following a measurement, are important for assessing how meaningful a measurement is. For example, increasing measurement durations does not always improve the accuracy of the results. Further, sampling more often with shorter intervals between the samples may not necessarily increase the amount of information.

The 802.11k draft standard does not specify default measurement durations. A station that requests a measurement can, however, specify its duration. Of course, if a measurement is performed without a previous request, the measuring station itself determines the duration.

---

* S. Mangold, Z. Zhong, K. Challapali, and C. T. Chou. Spectrum Agile Radio: Radio Resource Measurements for Opportunistic Spectrum Usage. In: 47th annual IEEE Global Telecommunications Conference, Globecom 2004, Dallas TX, USA, 29 November–3 December 2004. © 2004 IEEE

Similarly to the decision about the actual time when the measurement request is issued, and eventually to the decision about the interval after which requests may be repeated, it is a local decision of the requesting station to determine how long to measure, and whether or not to repeat it after a certain time.

The local time correlation of the medium usage (for example, the pattern of busy and idle times) is important information for optimizing such parameters, as discussed by Mangold *et al.* (2004a). What the optimal measurement parameters are, and how long reported results are valid and relevant for future events, generally depends on the scenario and types of radio system that operate in the medium, as well as on the offered traffic (e.g. packet sizes, arrival rates).

The improvement in the stochastic confidence in the results of reported radio resource measurements as a new approach to improve reliability of spectrum opportunity identification has been described (Mangold and Berlemann, 2005).

4.4.1.1 Medium Sensing Time Histogram Details

One measurement that was finally not accepted as a part of 802.11k, but which is important for cognitive radio enablers is the medium sensing time histogram report. This report was developed by Zhong *et al.* (2003), and discussed in detail by Mangold and coworkers (Mangold *et al.*, 2004a,b). This measurement would enable spectrum opportunity identifications based on CCA and noise patterns. A measuring station reports the histogram of medium busy and idle time observed during the measurement duration. The states 'busy' and 'idle' are typically defined by the CCA process, but may vary depending on what the requesting station attempts to gain from the measurement. Similarly to the noise histogram, non-802.11 energy can be measured, which makes the medium sensing time histogram a powerful measurement for 802.11-based cognitive radios, where 802.11 stations would have to measure the interference patterns from other, non-802.11 radio networks.

The interval durations during which no transmission occurs, referred to as idle durations, depend obviously on the activities of the communicating radio stations. The more medium accesses occur, the shorter are the idle durations. This is an intrinsic characteristic of the collision avoidance protocol used in WLAN IEEE 802.11. Idle durations can be measured, and stochastically evaluated, with the help of the medium sensing time histogram. Specifically, the CCA idle report will indicate the duration of idle times of the medium. Figure 4.1(a) illustrates the structure of the report. A so-called density vector is used to report indicators (referred to as bins, which are represented by one byte, therefore the value for a density varies between 0 and 255) for the probability of occurrence of specific medium idle durations. The size of the vector (the number of bins) is variable and, for example, defined by a measurement request from another station. Indicated in the figure is a vector with six bins.

Which bin corresponds to which duration is also a variable parameter and may have been defined by a measurement request that was received prior to the measurement. Figure 4.1(a) illustrates an example with CCA idle durations that correspond to slots of the 802.11 medium access protocol (SIFS, PIFS, DIFS, etc.).

Figures 4.1(b–d) illustrate typical measurement results for this configuration. Figure 4.1(b) shows a typical result that would occur if the medium is relatively idle, when stations access the medium only from time to time. Longer idle durations would occur in this case, and no relevant tendency of the histogram can be expected (usually, randomly distributed idle times).

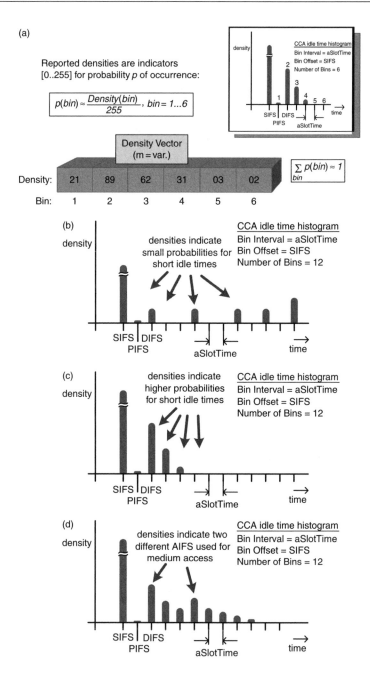

Figure 4.1 The medium sensing time histogram illustrated. Shown are typical results for idle durations, i.e. resulting histograms for durations of idle times between transmitted frames. Naturally, the SIFS duration occurs more often than other durations, because of the timing of frame exchanges in the 802.11 protocol. (a) CCA idle report with six bins. (b) CCA idle densities with some activities. (c) CCA idle densities with busy medium. (d) CCA idle densities with EDCA activity. Reproduced with permission from S. Mangold, Z. Zhong, K. Challapali, and C. T. Chou, Spectrum Agile Radio: Radio Resource Measurements for Opportunistic Spectrum Usage. In: 47th annual IEEE Global Telecommunications Conference, Globecom 2004, Dallas TX, USA, 29 November–3 December 2004. © 2004 IEEE

Figures 4.1(c) and (d) illustrate the resulting histogram of typical idle duration that would occur when the medium is very busy, and frame exchanges are separated by backoff processes only. In this case, as will be explained in the following chapters, the resulting histogram would show a clear tendency towards a specific distribution. The more busy the medium is, the higher is the probability of earlier medium access. Hence, earlier slots show higher probabilities of medium access (which means, in the language of 802.11k, a higher bin density). The difference between Figures 4.1(c) and (d) is that in (c), all stations operate according to the same EDCA parameter set, but in (d) two different EDCA priorities are used, and two different earliest medium access times (AIFS) can be identified.

4.4.2 IEEE 802.11n for High Throughput

IEEE 802.11n (IEEE, 2008b) is developing high throughput enhancements to 802.11 for multimedia applications and consumer electronics. The basic motivation of 802.11n is to combine the speed of fixed Ethernet with mobility from WLANs. Because of its promising economic potential, many interest groups are involved in 802.11n, leading to a complicated and lengthy standardization process. 802.11n was initiated in late 2003 and is expected to be finalized at the end of 2009. A maximum data rate of 600 Mb/s is targeted while today's 'draft n' devices in practice support 300 Mb/s (with two spatial streams at 40 MHz).

In order to increase the data rate of WLANs, 802.11n introduces enhancements to the physical layer as well as to the medium access control layer as described below.

The throughput enhancements of 802.11n at the physical layer are based on Multiple Input Multiple Output (MIMO) operation and channel bonding for 40 MHz operation. The PHY enhancements of 802.11n for increasing the data rate are limited by the 802.11 MAC protocol. 802.11n introduces, therefore, several means of reducing the overhead of the 802.11 legacy MAC protocol. Frame aggregation and block acknowledgements provide a further improvement in performance. Additionally, the interframe spacing is reduced by replacing the Short Interframe Space (SIFS) through the Reduced Interframe Space (RIFS). A Reverse Direction (RD) protocol and the Phased Coexistence Operation (PCO) are optional features that further improve the legacy 802.11 MAC protocol.

4.4.2.1 Multiple Input Multiple Output

MIMO uses multiple antennas for transmission and reception in order to improve the overall system performance. In general, there are three MIMO categories: spatial multiplexing, diversity and beamforming/precoding.

Diversity exploits independent fading in multiple antenna links to enhance the multipath signal diversity. A single stream (unlike multiple streams in spatial multiplexing) is transmitted in using Space–Time Block Coding (STBC). The multipath signals resulting from reflection during signal propagation arrive after the direct, line-of-sight signal at the receiver. Diversity increases the receiver's ability to recover information from a received signal. For applying diversity, no channel knowledge at the transmitter is required.

Beamforming/precoding increases the signal gain from constructive combining and reduces multipath fading. The same signal is emitted from each transmit antenna with appropriate phase weighting, such that signal power is maximized. For applying beamforming/precoding, knowledge of Channel State Information (CSI) at the transmitter is required.

Parallel channels are created in spatial multiplexing, where so-called streams transmit independent and separately encoded data signals from each of the multiple transmit antennas. Thus, independent data streams are simultaneously transmitted within one channel. Each stream requires an independent antenna at the transmitter and receiver. Additionally, separated radio frequency chains and analog-to-digital converters are required for each MIMO antenna, which results in a substantial increase in implementation costs. Spatial multiplexing is good at a higher Signal to Noise Ratio (SNR) and can be used with or without transmit channel knowledge.

4.4.2.2 Channel Bonding and Operation with 40 MHz instead of 20 MHz Bandwidth

The concept of channel bonding refers to adapting the channel bandwidth, for example from 20 to 40 MHz, dynamically. This is of high relevance to dynamic spectrum access, as it can be used to morph into various spectrum opportunities as part of the dynamic spectrum access. Therefore, we see 802.11n as an important enabler for cognitive radio. 802.11n applies channel bonding to allow use of a pair of adjacent 20 MHz channels as one 40 MHz channel. With a 40-MHz channel bandwidth, more data can be transmitted in less time, thus reducing the frame durations. Operation with a 40-MHz channel bandwidth is implemented and controlled by the access point in defining a primary and secondary 20-MHz channel. The primary 20-MHz channel is used for coordination (beacon broadcast, etc.) and stations do not need to consider the carrier sense state of the secondary channel when transmitting at 40 MHz. Additionally, stations shall not change their transmission time of a frame due to CSMA/CA on secondary channel. However, channel bonding has a main problem: potential collisions with 20-MHz transmissions from overlapping BSS on a secondary channel might occur leading to an interference in frame sequence at 40 MHz on the secondary channel.

The 802.11n 40-MHz channels are still under political discussion. The Bluetooth interest groups, for example, fear an essential threat to their Bluetooth technology with frequency hopping in case several 802.11n 40-MHz transmissions block the complete unlicensed spectrum at 2.4 GHz. The current draft of 802.11n intends to introduce a self-limitation on 20-MHz operation in case legacy 802.11 devices are detected. Additionally, any coexisting device independent of the transmission technology can use a '40 MHz intolerant' bit to request a deactivation of the 40-MHz operation. To generate this '40 MHz intolerant' bit, a non-802.11 device has, nevertheless, to implement some 802.11 functionalities, which implies additional device/technology complexity and thus, additional costs. Anyway, 802.11n devices are searching in both 20-MHz channels for ongoing transmission before initiating a 40-MHz operation.

4.4.2.3 Frame Aggregation

Similarly to the QoS enhancements of the base standard originating in 802.11e, 802.11n can aggregate multiple frames for reducing the protocol overhead. Two types of frame aggregation exist: first, the aggregation of MAC Service Data Units (MSDUs) at the top of the MAC (referred to as MSDU aggregation or A-MSDU), and second, the aggregation of MAC Protocol Data Units (MPDUs) at the bottom of the MAC (referred to as MPDU aggregation or A-MPDU). Figure 4.2 illustrates the first type in aggregating several MSDUs to an A-MSDU. The aggregated MSDU frames need to have the same address type and priority class.

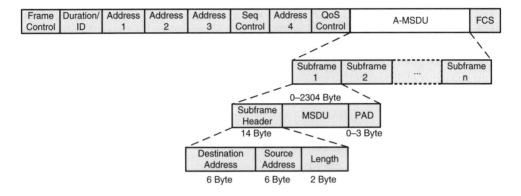

Figure 4.2 Frame aggregation of 802.11n. An A-MSDU is aggregated with multiple MSDUs

4.4.2.4 Block Acknowledgements

Block acknowledgements allow the receiving station to indicate a successful reception of more than one frame transmission with a single message. The basic functions of the block acknowledgement are defined within 802.11e and are, therefore, now part of the 802.11 base standard (IEEE, 2007). This basic principle of block acknowledgements is illustrated in the MAC timing diagram of Figure 4.3.

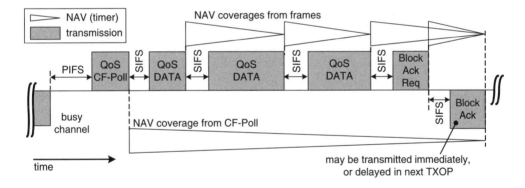

Figure 4.3 Timing diagram of block acknowledgement (Walke *et al.* 2006)

802.11n extends the block acknowledgement concept from 802.11e in several aspects: Implicit Block Acknowledgement Request (BAR) and multi-traffic identifier block acknowledgement help to reduce the protocol overhead. Furthermore, a block acknowledgment may be immediate or delayed, depending on the traffic type.

4.4.2.5 Reverse Direction Protocol

The Reverse Direction (RD) protocol defines two roles, the RD initiator and RD responder. With the RD protocol, the RD initiator may transmit PPDUs and directly obtain response

PPDUs from the RD responder during frame exchange. In this way, the response message can be received directly without requiring additional successful contention-based channel access from the RD responder. Thus, response time, channel idle time and collision probability are reduced. A so-called Reverse Direction Grant (RDG) is used for indicating a request for usage of the RD protocol. Due to traffic requirements or limited capabilities the RD responder might not be able to use the RDG and is therefore not required to use this grant.

4.4.2.6 Phased Coexistence Operation

The high throughput features of IEEE 802.11n also imply, among other things, an (optional) cognitive radio feature, namely the Phased Coexistence Operation (PCO). It is designed to allow an 802.11n access point to support 802.11n and 802.11a (and b/g) clients. For this, PCO implements a phased operation with flexible channel bandwidths of 20 and 40 MHz. Two neighboring 802.11 channels, each with a 20 MHz bandwidth, are categorized as a primary and a secondary channel. With PCO, the 802.11n access point is able to switch between two-times 20-MHz and 40-MHz phases of operation. Thus, with PCA an 802.11n access point can support and control legacy devices in the two 20-MHz channels as well as benefiting from 40-MHz operation when communicating with 802.11n stations. The PCO of 802.11n is an initial step towards multi-channel operation with flexible bandwidth usage, here centrally coordinated on the level of the MAC layer.

The PCO of 802.11n is illustrated in the MAC timing diagram in Figure 4.4. Two orthogonal neighboring channels, each of 20-MHz bandwidth are depicted. Channel f_1 is marked as the primary channel through a respective beacon transmission, while channel f_2 is regarded as secondary channel. Additionally, the Network Allocation Vectors (NAVs) of two legacy 802.11 (20 MHz) stations operating on f_1 and f_2 respectively are shown. In the last row, the combined channel view of an 802.11n (40 MHz) station is depicted. A centrally operating 802.11n access point is setting the NAVs of all stations with adequate control frames transmitted on the respective channel for channel reservation. The access point is thus coordinating the phased operation of the associated stations with 20- and 40-MHz bandwidth usage.

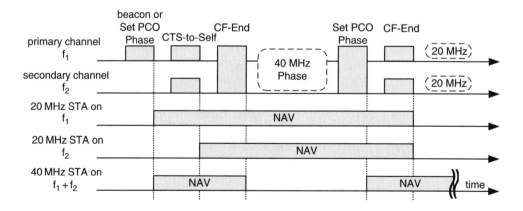

Figure 4.4 Timing diagram for the (optional) 802.11n Phases Coexistence Operation (PCO)

4.4.2.7 Wi-Fi Certified 802.11n Draft 2.0

In mid-2007, the Wi-Fi Alliance initiated the certification of products based on the IEEE 802.11n Draft 2.0 standard. This certification program established a *de-facto* standard for 'draft n' devices in defining a set of features and a level of interoperability across different vendors. Table 4.2 summarizes the respective features and quantifies the expected throughput performance enhancement (Wi-Fi-Alliance, 2008). Many commercial products of various vendors have been certified according to this feature set and are allowed to use the Wi-Fi CERTIFIED 'draft n' logo.

Table 4.2   Wi-Fi Certified 802.11n Draft 2.0 feature set (Wi-Fi-Alliance, 2008)

Feature	Explanation	Type	Potential throughput enhancement
Support for two spatial streams in transmit mode	Required for an AP device	Mandatory	100% → from MIMO PHY
Support for two spatial streams in receive mode	Required for an AP and a client device, except for handheld devices	Mandatory	
Support for A-MPDU and A-MSDU	Required for all devices	Mandatory	<100% (depending on traffic characteristics) →from MAC
Support for block ACK	Required for all devices	Mandatory	
2.4 GHz operation	Devices can be 2.4 GHz only, 5 GHz only or dual-band.	Tested if Implemented	
5 GHz operation	Devices can be 2.4 GHz only, 5 GHz only or dual-band.	Tested if Implemented	
40 MHz channels in the 5 GHz band	40 MHz operation is only supported by the Wi-Fi Alliance in the 5 GHz band. 2.4 GHz may be certified later.	Tested if Implemented	100% → from MIMO PHY
Greenfield preamble	Greenfield preamble cannot be interpreted by legacy stations. It is shorter than the mixed mode or legacy mode preamble and improves efficiency of the 802.11n networks with no legacy devices.	Tested if Implemented	
Short Guard Interval (Short GI), 20 and 40 MHz	Short GI is 400 nanoseconds vs. the traditional GI of 800 nanoseconds. Short GI reduces the symbol time from 4 microseconds to 3.6.	Tested if Implemented	10% → from MIMO PHY
Concurrent operation in 2.4 and 5 GHz bands	This mode is tested for access points only	Tested if Implemented	

4.4.3 IEEE 802.11s for Mesh Networks

Since 2004, the 802.11s task group has been developing a mesh extension for IEEE 802.11 (IEEE, 2008c). The scope is limited to enhancements for mesh networking on the MAC layer without requiring a redesign of the PHY layer. An architecture and protocol for implementing small/medium mesh networks (~32 nodes) with self-configuring, multi-hop topologies is standardized.

The basis of 802.11s was laid in July 2005 when 15 proposals were received in response to a call for proposals issued by 802.11s. Subsequently, two competing industry consortia (the Wi-Mesh Alliance and the SEE-Mesh Consortium) were formed. Their system concepts were merged in March 2006 to form the first baseline 802.11s draft amendment resulting to the following four main pillars of 802.11s:

- interworking,
- medium access control,
- security, and
- path selection.

Further details on 802.11s can be found in the literature (Hiertz *et al.*, 2008a,b). Spectrum sharing in IEEE 802.11s wireless mesh networks has also been evaluated (Max *et al.*, 2006), with the well-known IEEE 802.11 DCF being compared to a distributed, reservation-based approach from the Mesh Network Alliance (MNA). The contention-based distributed medium access of 802.11 and its extension for usage in multiple parallel-frequency channels is highlighted by Mangold and Habetha (2005). A multi-channel station groups channels to increase its own achievable throughput. Additionally, methods to enable the coexistence of multi-channel stations and single channel are introduced.

4.4.3.1 Network Topology

802.11s Wireless Mesh Network (WMN) terms and definitions are illustrated in Figure 4.5. The basic entity of an 802.11 network is station implemented through an 802.11-standard-compliant MAC and PHY (physical layer). Two stations from an elementary 802.11 network are referred to as a Basic Service Set (BSS). If a station provides an integration service to other stations, it is referred to as an access point and an infrastructure BSS is formed. A so-called Distribution System (DS) provides the services that are necessary to communicate with devices outside the station's own BSS. Multiple, interconnected, BSSs establish an Extended Service Set (ESS). Within an ESS, stations can roam from one BSS to another. Today's Ethernet (802.3) usually provides the Distribution System Medium (DSM) that the DS relies upon. The entity that provides the integration of the WLAN with non-802.11 networks is referred to as a 'portal' and access points are collocated with that portal.

A mesh point is the basic mesh device. Several mesh points can form a WLAN mesh by establishing mesh links with neighboring mesh points. Depending on collocated functionality, a mesh point may integrate the access point (referred to as mesh access point), portal, or other services. Interoperability between the different networking concepts is a requirement for the 802 family of standards. Therefore, the 802.11s network appears as a single Ethernet

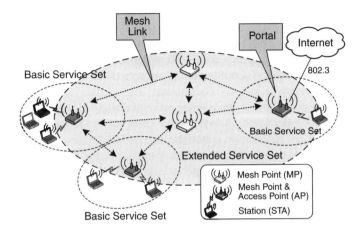

Figure 4.5 Topology of a 802.11s wireless mesh network

segment to the outside, allowing seamless integration. The WMN implements a single broadcast domain and thus integrates seamlessly with other 802 networks. In particular, 802.11s supports transparent delivery of uni-, multi- and broadcast frames to destinations inside and outside the mesh. A mesh point with integrated portal functionality is referred to as a Mesh Portal Point (MPP).

4.4.3.2 Frame Concept

802.11s extends the 802.11 MAC frame categories for data, control or management for multi-hop operation. To provide for multi-hop, 802.11s extends the 802.11 frame by an additional 'mesh header' as shown in Figure 4.6. The mesh header consists of a mesh time to live field, a mesh sequence number, a mesh flags field, and an optional mesh address extension field. The mesh time to live and sequence number fields are used to prevent infinite frame loops. The mesh flags field indicates the presence of additional MAC addresses in the mesh header, and the optional address extension allows up to six address additional fields in a mesh frame, which is useful when the source and destination of the frame are not part of the mesh, but are proxied by mesh points.

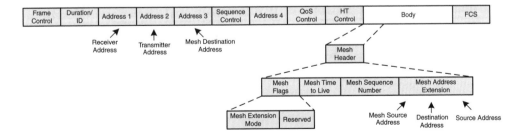

Figure 4.6 IEEE 802.11s mesh header as part of the frame body

4.4.3.3 Medium Access Control Protocol

Mesh points use the Enhanced Distributed Channel Access (EDCA) of the base standard (IEEE, 2007) for their medium access. This contention-based medium access is an enhancement of the Distributed Coordination Function (DCF), as described in detail in Section 5.1 in the context of WLAN coexistence. The medium access relies on Carrier Sense (CS) and does not require synchronization between mesh points. Before an 802.11 station transmits, it uses the Clear Channel Assessment (CCA) to ensure that the wireless medium is idle. With EDCA, a station may transmit multiple frames whereas the total transmission duration may not exceed the so-called transmission opportunity limit. The intended receiver acknowledges any successful frame reception. Additionally, EDCA differentiates four traffic categories with different priorities in medium access and thereby allows for limited (probabilistic) support of Quality of Service (QoS).

Originally, EDCA was been designed for a single-hop topology. Its application in WMNs therefore has several performance drawbacks. While in a single-hop wireless network, all devices are either in mutual reception range or have at least one common neighbor device, WMN can be regarded as continuously overlapping neighboring networks. Thus, devices are mutually unaware of each other. The resulting hidden and exposed node problems lead to mutual interference and inefficient retransmissions.

In many cases, the WMN replaces a wired infrastructure of interconnecting multiple access points. Thus, it has to transport the aggregated traffic of all devices. The 802.11s base standard is not able adequately to prioritize the backhaul traffic and provide mutual negotiation and coordination between neighboring access points of the WMN.

To overcome these performance limitations, 802.11s provides an optional MAC scheme referred to as Mesh Deterministic Access (MDA). MDA as part of the current 802.11s draft standard (IEEE, 2008c) is a joint compromise of the competing Wi-Mesh Alliance and SEE-Mesh consortium. The MDA divides the mesh wide super frame into slots of 32-μs duration as illustrated in Figure 4.7. An amount of subintervals within the superframe is defined by the periodicity of an MDA reservation. A reservation within each subinterval is defined by an offset value and a corresponding duration. Such a reservation is referred to as an MDA Opportunity (MDAOP). A MDAOP grants its owner highest priority for medium access as reflected, for example, by a zero backoff for accessing the wireless medium. Within an MDAOP, the frame exchange is equivalent to a frame exchange during a transmission opportunity and any

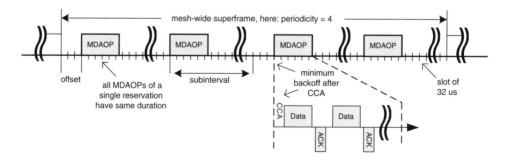

Figure 4.7 MAC timing diagram of the (optional) 802.11s mesh deterministic access (MDA)

kind of acknowledgment scheme may be used. Neighboring MDA-capable devices defer from medium access during the MDAOP, similarly to the transmission opportunity of 802.11. To set up an MDA reservation, the initiating mesh point uses an MDAOP set-up request message. Depending on its local knowledge about neighboring MDAOPs, the receiving device accepts the set-up request. Both mesh points include the MDAOP in their beacon frame and send reservation details with unicast frames to neighbors if requested. Mesh points advertise an interfering times report to neighbor mesh points. This enables within the extended neighborhood of mesh points knowledge of slots that are currently in use. The fraction of the superframe that can be used for MDA is defined by the MDA fraction (MDAF) parameter broadcast in the beacon. An MDA-capable device can reject additional MDAOP reservations in the case of this limit being exceeded.

The MDA replaces the contention-based medium access of the EDCA through reservations. Nevertheless, no performance or QoS guarantee is possible as MDA-unaware devices in the same frequency channel can delay or disable MDA reservations.

4.4.3.4 Congestion Control

The medium access of 802.11 relies on carrier sensing. A mesh point located at the edge of a WMN has fewer neighbors and therefore senses an idle medium more often than do mesh points in the center of the WMN. Thus, mesh points at the edge have a higher probability to transmit. In case of congestion within the WMN center, frames that have been relayed over several hops may be dropped. The optional 802.11s congestion control concept uses a management frame to indicate the expected duration of a congestion. Each node actively monitors local channel utilization. If congestion is detected, previous-hop neighbors are notified and a neighbor mesh point can be requested to slow down. Local rate control (and signaling) is done on a per-access category basis, e.g. the data traffic rate may be adjusted without affecting voice traffic.

4.4.3.5 Routing Path Selection in Multi-hop Scenario

Wireless routing on layer 2 (MAC) is referred to as path selection to differentiate it from routing on layer 3 (IP). 802.11s defines an extensible framework that supports mandatory and alternative path metrics and path selection protocols.

The default path metric is referred to as 'airtime metric' and indicates overall cost of a link by taking into account data rate, overhead and frame error rate of a test frame with a size of 1 kilobyte. The default path selection protocol is called Hybrid Wireless Mesh Protocol (HWMP): It combines on-demand, distributed route discovery (flexible in changing environments) with proactive, tree-oriented routing (efficient in mesh networking). The on-demand routing is based on the Radio Metric Ad Hoc On Demand Distance Vector (AODV) (RM-AODV) protocol and allows destinations to be discovered in the mesh on demand. In the presence of a root portal, a distance vector routing tree is built and maintained and the portal may establish and maintain paths to all mesh points.

4.4.3.6 Common Channel Framework

The primary aim of 802.11s, the creation of a wireless distribution system with automatic topology learning and wireless path configuration, implies many cognitive radio aspects. An awareness of the local communication environment is required in order to allow a dynamic, radio-aware, path selection within the mesh.

An optional feature of 802.11s, namely the Common Channel Framework (CCF), introduces a coordination channel and the corresponding MAC protocol enhancements for using different orthogonal frequency channels. Contrary to the PCO of 802.11n, these channels are not required to be neighboring. With CCF, stations periodically switch to the common channel for distributed channel reservation through suggestion and acceptance of destination channels.

The principle of the multi channel MAC protocol implemented through the CCF is illustrated in Figure 4.8 with a MAC timing diagram. Channel f_0 is used as common coordination channel. On this common channel f_0, 802.11s stations coordinate their data transmissions on the data channels f_1 to f_n. This is done in a dedicated phase, referred to as a channel coordination window, with Request-to-Switch (RTX) and Clear-to-Switch (CTX) messages. The channel coordination window has a repetition period, which enables, as shown, normal data transmissions outside this window on the common channel.

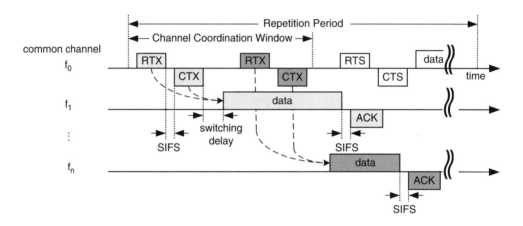

Figure 4.8   MAC timing diagram of the (optional) 802.11s common channel framework (CCF)

4.4.4 IEEE 802.11y for High Power Wi-Fi

IEEE 802.11y is working towards a standard for high power Wi-Fi equipment operating in the 50-MHz frequency band at 3.65–3.70 GHz (IEEE, 2008a) and standardization is intended to be finalized in September 2008. The 3.650–3.700 GHz frequency band has been opened to unlicensed services (FCC, 2004b). A contention-based, listen-before-talk, protocol is demanded there in order to operate in this frequency band. Therefore, the FCC has defined (FCC, 2005a) a contention based protocol as:

'A protocol that allows multiple users to share the same spectrum by defining the events that must occur when two or more transmitters attempt simultaneously to access the same channel and establishing rules by which a transmitter provides reasonable opportunities for other transmitters to operate. Such a protocol may consist of procedures for initiating new transmissions, procedures for determining the state of the channel (available or unavailable), and procedures for managing retransmissions in the event of a busy channel.' (FCC, 2005a)

Some of these characteristics are satisfied by currently available 802.11a systems operating in the U-NII band at 5 GHz. These are frequency agile, have the ability to sense signals from neighboring transmitters, offer adaptive modulation, and use transmit power control.

The FCC issued final rules for a novel 'light licensing' scheme in June 2007 for the 3650–3700 MHz band (FCC, 2007). A small fee for a nationwide, nonexclusive, license is required. Additionally, a nominal fee for each high powered Base Station (BS) that is deployed has to be paid. Contention-based protocols are mandatory so as to allow fair and opportunistic spectrum sharing in case of interference between the licensee's devices.

802.11y introduces a combination of higher power limits and enhancements made to the PHY/MAC timings of the 802.11 base standard that allow Wi-Fi devices to operate at distances of more than 5 km. Additionally, a flexible spectrum management framework of regulatory classes and behavior limit sets is defined for adopting coverage and spectrum

usage according to local regulation requirements. Exemplary regulatory classes and behavior-limits sets that are used by regulatory classes are shown in Table 4.3.

Table 4.3 Exemplary behavior limits sets and regulatory classes (IEEE, 2008a)

Behavior limits set	Topic	USA	Europe	Japan
3	Transmit power control	Reserved	ETSI EN 301 389-1	Reserved
4	Dynamic frequency selection	Reserved	ETSI EN 301 389-1	Reserved
6	4 ms Carrier Sensing	4 ms, no exceptions	Reserved	MIC EO Articles 49.20, 49.21
9	Public safety	FCC 47 CFR [B8], Section 90.1209	Reserved	Reserved
10	Part 15 license exempt bands	FCC 47 CFR [B8], Section 15.247	ETS 300-328	MIC EO Article 49.20

Regulatory class	Channel starting frequency (GHz)	Channel spacing [MHz]	Channel set	Transmit power limit (mW)	Transmit power limit (EIRP)	Emissions limits set	Behavior limits set
10	4.85	20	21, 25	100	—	5	9
11	4.85	20	21, 25	2000	—	5	9
12	2.407	25	1-11	1000	—	4	10
13	3.000	20	133, 137	—	1 W/MHz	6	3, 4, 6, 11, 15
13	3.000	20	133, 137	—	40 mW/MHz	6	3, 4, 5, 6, 12, 15
14	3.000	10	132, 134, 136, 138	—	1 W/MHz	6	3, 4, 6, 11, 15
14	3.000	10	132, 134, 136, 138	—	40 mW/MHz	6	3, 4, 5, 6, 12, 15

Further, 802.11y is introducing smaller channel bandwidths to Wi-Fi as it is capable of operating with 5-, 10- and 20-MHz channel bandwidth. The 802.11y Extended Channel Switch Announcement (ECSA) is an enhanced Dynamic Frequency Selection (DFS) procedure and provides the means for notifying stations of changing channels and channel bandwidth. An operator can extend and retract permissions to license exempt devices (referred to as dependent stations in 802.11y) to use licensed radio spectrum by applying Dependent Station Enabling (DSE). This mechanism guarantees an operation in licensed spectrum depending on approval by periodic messages received from the operator's BS. Further, DSE can be extended to channel management coordination. A location-dependent spectrum access and interference resolution is implemented, based on the location of the access granting BS. Thus no locating technologies like GPS are required in the 802.11y station (price, device complexity and indoor failure are reduced).

The light licensing concept of the 802.11y spectrum management framework is not limited to operation at 3.6 GHz; instead it provides the general means for Wi-Fi to be operated at any spectrum, such as, for example, the candidate bands for IMT-advanced like 450–862 MHz or 2300–2400 MHz. As consequence, 802.11y is also a promising competitor of 802.22 for the reuse of TV bands.

4.5 WiMAX Wireless Metropolitan Area Networks IEEE 802.16

4.5.1 IEEE 802.16.2 Coexistence

Coexistence in Broadband Wireless Access (BWA) Systems of IEEE 802.16 operating in licensed bands was standardized by the task group IEEE 802.16.2. This task group provided a recommended practice for the design and coordinated deployment of BWA systems in order to control interference and facilitate coexistence between these systems and other applicable systems that may be present. It analyzes appropriate coexistence scenarios and provides guidance for systems design, deployment, coordination, and frequency usage. It generally addresses licensed spectrum between 2 GHz and 66 GHz.

The IEEE standard 802.16.2–2004 was published in 2004 (IEEE, 2004b). 802.16.2 suggests threshold parameters, like the distance between two interfering base stations, to assess the necessity for inter-operator coordination. These parameters are used to define guidelines for geographical spacing and frequency reuse. A concept of using Power Spectral Flux Density (PSFD) values is introduced in order to trigger different levels of initiatives taken by an operator to give notifications to other operators. Maximum values that can be tolerated as a result of co-channel interference originating from adjacent operators are defined for the PSFD. Frequency guard bands, recognition of cross-polarization differences, antenna angular discrimination, spatial location differences, the use of adaptive antennas, and frequency assignment substitution, are suggested to reduce the probability of interference.

4.5.2 IEEE 802.16h License Exempt

The IEEE 802.16h License-Exempt Task Group is developing improved mechanisms for enabling coexistence between license-exempt systems based on IEEE 802.16. The standardization efforts target, for example, MAC enhancements and policies. Additionally, 802.16h also focuses on the coexistence of such systems with primary radio systems when hierarchically sharing spectrum. The amendment is not limited to license-exempt bands, but operation may take place in all bands where 802.16–2004 is applicable.

A distributed architecture for radio resource management has been suggested (IEEE, 2005a), which enables communication and exchange of parameters between different networks formed by one 802.16 BS and its associated subscriber stations. Each BS has a Distributed Radio Resource Management (DRRM) entity to execute the spectrum-sharing policies of 802.16h and to build up a database for sharing information related to actual and intended future usage of radio spectrum.

802.16h proposes a coexistence protocol to realize all functions required for coexistence such as, for example, detecting the neighborhood topology, registering to the database, or negotiation for sharing radio spectrum. The DRRM uses the coexistence protocol to communicate with other BSs and regional license-exempt databases while interacting with MAC or PHY.

4.5.3 IEEE 802.22 for Wireless Rural Area Networks

From its organization, IEEE 802.22 is separated from WiMAX IEEE 802.16 and the project group works independently. However, we find many similarities in the candidate technology selected for 802.22, such as transmission schemes and the medium access protocol. This is an important fact that may be interpreted as 802.22 standardization, having selected a relative conservative approach that could be seen as less ambitious as compared with P1900, and also less ambitious compared with what researchers on cognitive radio are trying to realize. We decided therefore to place our section on IEEE 802.22 under the umbrella of IEEE 802.16, to reflect this situation.

The IEEE 802.22 working group is targeting the standardization of a cognitive air interface for fixed, point-to-multipoint, Wireless Regional Area Networks (WRANs) operating on unused channels in the VHF/UHF TV bands between 54 and 862 MHz. Thereby, 802.22 will provide wireless broadband access from a BS to rural areas over distances of typically 15 to 30 km and serving up to 255 fixed Customer Premise Equipments (CPEs). The technical basis of 802.22 is WiMAX, with the following key characteristics: OFDMA, TDD, a 10-ms MAC frame, channel bandwidth of 6, 7, 8 MHz and a peak data rate of 72.6 Mb/s (Mbps) (with optional channel bonding and channel aggregation).

In order to apply 802.22 in Europe, the TV signal detection and channel bandwidths have to be modified and adapted to the European TV broadcast standards.

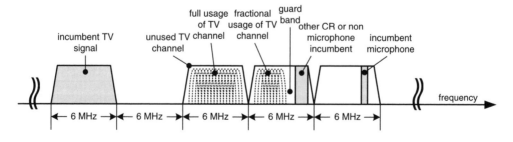

Figure 4.9 Diagram of fractional bandwidth usage of 802.22 in the frequency domain (IEEE, 2006). Five TV channels each of 6 MHz are depicted

4.5.3.1 Protection of Incumbents

Geographic databases and location technologies are not part of 802.22. Instead, incumbent awareness and interference prevention are realized as part of MAC and PHY through the following:

- distributed spectrum sensing and spectrum management,
- quiet period and fast/fine sensing management,
- measurements and clustering,
- detection algorithms.

All 802.22 devices (BS and CPE) sense the spectrum for three different licensed transmissions: (i) analog television, (ii) digital television, and (iii) licensed low power auxiliary devices, such as wireless microphones operated by TV broadcast service providers.

A so-called spectrum manager implements cognitive radio functions at the BS. It will use inputs from a Spectrum Sensing Function (SSF), geolocation and an incumbent database to decide on the TV channel usage as well as the EIRP limits imposed to specific 802.22 CPEs.

4.5.3.2 Fractional Bandwidth Usage

The fractional bandwidth usage of 802.22 implements a dynamic spectrum access (i.e. adaptive channel allocation) according to the incumbent's usage of the channel. Figure 4.10 illustrates the fractional bandwidth usage of 802.22 in the frequency domain (IEEE, 2006). The fractionally vacant bandwidth of a single channel can be used by consideration of location, bandwidth and type of narrowband incumbent users. Thereby, the number of subcarriers used by 802.22 is proportional to the fractional bandwidth. In case wireless microphones are operating the co-channel, 802.22 systems shall free the entire channel. A guard band is used to protect other cognitive radio or non-microphone incumbent radio transmissions. With fractional bandwidth usage, the 802.22 channel bandwidth can be extended over two incumbent TV channels. The fractional BW mode is identified by using a preamble.

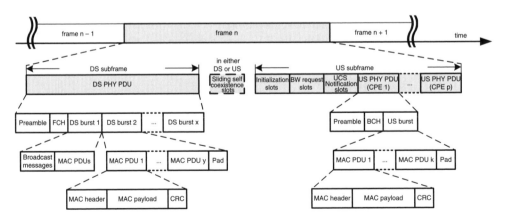

Figure 4.10   Timing diagram of the 802.22 MAC frame in the time domain

4.5.3.3 Medium Access Control

The MAC frame of 802.22 is illustrated in Figures 4.10 and 4.11. A MAC frame in the time domain is shown in Figure 4.10, while Figure 4.11 depicts the MAC frame and the higher level of superframe in the time and frequency domain.

Each 802.22 MAC frame is divided into a Downstream (DS) subframe and an Upstream (US) subframe with an adaptive boundary in between. The DS subframe contains a single PHY PDU addressed to multiple 802.22 CPEs. However, the US subframe may contain

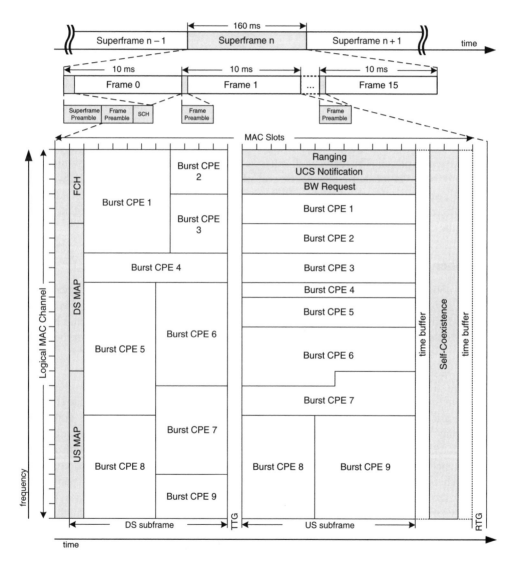

Figure 4.11 MAC frame composition of 802.22. The time and frequency domain are divided into MAC slots and logical MAC channels. In this way the OFDMA of 802.22 is realized

multiple PHY PDUs from various CPEs. Each PHY PDU is initiated with a preamble for synchronization and channel estimation. The preamble has a long training sequence and optional short training sequence. In the DS subframe, the Frame Control Header (FCH) contains information elements on size of DS and US subframe composition and channel descriptors for their PHY characteristics. The US subframe begins with contention intervals for initialization, bandwidth request and Urgent Coexistence Situation (UCS) notifications that precede the US PHY PDUs. The initialization slots are used for ranging in adapting the timing offset and power adjustments. The UCS notification slots are used for transmitting incumbent detection reports. The Burst Control Header (BCH), as part of the first US PHY PDU, contains the MAC addresses of BS and CPE in each US burst to enable BS/CPE identification from each US burst.

The 802.22 MAC superframe structure in the time and frequency domain is illustrated in Figure 4.11, which shows as an example the OFDMA implementation of 802.22. As illustrated, the 802.22 Superframe has a duration of 160 ms and consists of 16 MAC frames of 10-ms duration. The burst sizes are adapted under centralized coordination from the BS in the time and frequency domain according to their traffic requirements.

A Coexistence Beacon Protocol (CBP) based on beacon transmissions between the spectrum sharing WRAN cells implements decentralized coexistence coordination in the self-coexistence phase. The dedicated self-coexistence phase, consisting of sliding self-coexistence slots, is located either at the end of each 802.22 MAC frame or between the DS and US subframe. It is used by the CPE to improve coexistence with neighbors. Additionally, centralized coordination via the BS is implemented with the BS beacon being part of the Superframe Control Header (SCH).

4.6 Other Standardization Activities

A strong coordination between academia, regulation, infrastructure suppliers and operators is required in order to guarantee wide acceptance and commercial success of cognitive radio. The standardization of cognitive radio is an essential part of this coordination that helps to channelize the joined, interdisciplinary research and development activities all over the world.

4.6.1 White Spaces Coalition and Wireless Innovation Alliance

The White Spaces Coalition (founded in 2006) and the Wireless Innovation Alliance (founded in 2008) are both planning to provide high speed broadband Internet access to consumers via existing 'white spaces' in unused analog television frequencies, i.e. the digital dividend. These groups intend to challenge incumbent telephone and cable companies with a fixed, long-distance wireless broadband technology. Dell, Google, HP, Microsoft, and Motorola are among others participating in both groups while, for example, Intel, Philips, Samsung and Earthlink are limiting their involvement to the White Spaces Coalition.

Scientific publications and strong attendance of researchers allows the conclusion that these two groups will apply Wi-Fi as base technology for secondary access to TV bands. In this context, 802.11y from above forms an adequate basis for technical features.

Obviously, the time consuming and thus expensive process of IEEE standardization is avoided with the benefit of a short time-to-market. The goal seems to be an outrunning of 802.22 and coming up with commercial products for the opening of the digital dividend early in 2009. Microsoft, as a member of the White Spaces Coalition, submitted an initial proto-type to FCC for testing in 2007 and had to go through some cycles of resubmission due to objections by the FCC.

The dual-beaconing approach for realizing an operator-assisted cognitive radio, described later in this book, can be applied for implementing the reliable reuse of TV bands with enhanced Wi-Fi beacons. This is one of the key building blocks of the future of radio Inter-net developed by the European research project All-Wireless Mobile Network Architecture (WIP), see Section 4.1.5.

4.6.2 The New America Foundation and Open Spectrum

Open spectrum allows anyone to access any range of spectrum without permission under consideration of a minimum set of rules from the technical standards or etiquettes that are required for sharing spectrum (Weinberger *et al.*, 2005). Open spectrum targets the complete liberalization of radio communication in overcoming regulatory roadblocks. The core concept of open spectrum is that technologies and standards are able dynamically to manage the access and spectrum sharing, replacing the static spectrum assignments resulting from bureaucratic licensing (Berger, 2003).

New, upcoming radio transmission technologies suggest that radio spectrum can be treated as open access commons rather than a collection of strung-together access rights (Noam, 1998). It is right that just squeezing much more wireless communication out of a given band-width does not solve the tragedy of commons (Hazlett, 1998). However, rules and etiquettes on the one hand and technologies allowing more than one user simultaneously using the same frequency on the other, enable open spectrum. Thus, spectrum is different from other public resources, such as highways, that are overcrowded with too many cars.

Three primary technologies that have seen great development progress in recent years can be identified to realize open spectrum:

- low-power, UWB underlay spectrum usage (see, for example, Section 3.1.2);
- cognitive, frequency agile radios based on SDR (see Section 7.2);
- cooperation-based, self-configuring meshed.

As open spectrum considerably impacts the way spectrum is regulated, and regulation author-ities have to face new challenges resulting from this fundamental rethinking (Werbach, 2002):

- The scarcity of spectrum is overcome. Dynamic spectrum access through cooperative techniques improving efficiency makes regulatory limitation of spectrum access obsolete.
- The competition through innovation while sharing spectrum as a common resource is in many cases superior compared with auctioning licensed spectrum on the market.
- Investment costs for exploiting spectrum are essentially reduced as no tremendous fees for temporal licenses are required, resulting in essentially lower service costs.
- Wireless broadband technologies are a solution for the last-mile bottleneck. The spe-cial challenges of last-mile access can be solved by using long-range communications, wideband underlay or meshed architectures.
- Wide-area 3G systems can be complemented by more efficient short-range and meshed unlicensed technologies.

Noam (1997, 1998) approaches open spectrum under the term 'open access'. There 'open access' allows many users from different radio-based applications to access spectrum without license by buying access tickets, referred to as 'access tokens', whose prices depend on congestion. Transmitted data packets, piggy-backed, carry these electronic tokens and a decentralized payment is possible at tollgates, like, for example, access points. Such a system would reduce infrastructure costs to a marginal value, reduce market access barriers and encourage competition. Contrary to current unlicensed systems, which depend on etiquettes to manage competition, such a system would guarantee access under congestion conditions, as the individual value of spectrum is reflected in prices (Noam, 1997).

In the literature, open spectrum sometimes has a less revolutionary meaning in referring to the vertical sharing of licensed spectrum (Zhao *et al.*, 2005).

4.6.3 SDR Forum

The SDR (Software Defined Radio) Forum (SDRforum, 2005), founded in 1996, is a non-profit international industry organization promoting the development, deployment and use of SDR technologies for advanced wireless systems. The commercial pillar of the SDR Forum's strategy foresees the creation of standards and specifications that will reduce costs and time-to-market for SDR and cognitive-radio based systems and products. Third-party standards supported by the SDR Forum are explicitly wanted and own specifications and standards only intended when no one else can or will.

4.6.4 Third Generation Partnership Project 3GPP

Self-organizing networks have been recently taken up by 3GPP Release 8 in its TS32 series. With this step, 3GPP is approaching cognitive radio aspects from the cellular networks perspective. This work was done by the Technical Specification Group for Service and System Aspects in its Working Group 5 (TSG SA WG5) on 'Telecom Management'. TSG SA WG5 specifies a general management framework and requirements for enabling this management. Self-organizing networks concepts and requirements, self-establishment of eNodeBs, automatic neighbor relation management, as well as concepts and requirements for self-optimization and self-healing, are included in the ongoing standardization activities. Additionally, Working Group 3 of the Technical Specification Group for the Radio Access Network (TSG RAN WG3) is responsible for the overall architecture development and the specification of protocols is preparing a technical report on 'Self-configuring and Self-optimizing Network (SON) Use Cases and Solutions' for the next generation of UMTS referred to as Long Term Evolution (LTE).

The Self-Organizing Networks standardization activities of 3GPP are initiated by the Next Generation Mobile Networks (NGMN) Alliance (NGMN, 2007). NGMN represents 18 mobile network operators and was founded in 2006 in order to provide a coherent vision for the future mobile network technology evolution.

4.6.5 European Telecommunications Standards Institute ETSI

Cognitive radio can be based on SDR platforms, but this is not a necessity. The European Telecommunications Standards Institute (ETSI) regards these technologies as Reconfigurable

Radio Systems (RRS) exploiting reconfigurable radio/networks and self-adaptation to a dynamically changing environment, with the aim of ensuring end-to-end connectivity. ETSI recently established a technical committee on RRS with the following responsibilities:

- to study the feasibility of standardization activities related to RRS encompassing radio solutions related to SDR and cognitive radio research topics;
- to collect and define the related RRS requirements from relevant stakeholders;
- to identify gaps where existing ETSI standards do not fulfill the requirements, and suggest further standardization activities to fill those gaps.

The RRS technical committee shall address (from the perspective of ETSI so far uncovered) topics in standardization such as improved spectral utilization and inter-operator coexistence, using flexible usage modes and a variety of different Radio Access Technologies (RATs). As cognitive radio is related to national security aspects (for example, flexible encryption algorithms) this European driven standardization effort might additionally have strategic motives for the EU member states to be independent of the US-dominated IEEE.

4.6.6 Academic Research Conferences and Workshops

Since 2005, several periodic conferences and workshops (usually in conjunction with well-established academic research conferences) have been started, focusing entirely on the topic of cognitive radio and dynamic spectrum access. Among others, the following are of significance at the time this text was written. In particular, IEEE DySPAN can be highlighted as being the premier forum for discussion of all aspects of devices and networks that utilize spectrum on a dynamic basis.

- IEEE symposia on what is referred to as New Frontiers in Dynamic Access Spectrum Access Networks (DySPAN).
- Cognitive Radio Oriented Wireless Networks and Communications (Crowncom).
- SDR Forum Technical Conference.
- IEEE CogNet Workshop, in conjunction with ICC.
- IEEE International Workshop on Cognitive Radio Networks, in conjunction with CCNC.
- In Germany, the Karlsruhe Workshop on Software Radios (WSR).
- Periodic workshop on Cognitive Radio and Advanced Spectrum Management.

5

Proposed Enablers for Realizing Horizontal Spectrum Sharing

Coexistence and horizontal spectrum sharing are an increasingly significant technical problem and tend to constrain the future success of wireless communication. From a regulation perspective, equally righted transmission technologies are competing more and more to use the same limited spectrum. The unlicensed bands at 2.4 GHz are an example of this; for instance, Wi-Fi and Bluetooth have to operate in the same frequency bands and are victims of their own commercial success. Mutual interference due to devices operating in the same spectrum and location is already today a common occurrence. Additional horizontal spectrum sharing scenarios might arise in the near future as a result of the trend for technology-neutral spectrum regulation. A set of technology independent rules for spectrum usage, i.e. spectrum etiquettes, replaces the requirement to use a specific radio access technology in combination with a nonexclusive spectrum license.

This chapter proposes several enablers for realizing horizontal spectrum sharing, starting with today's technologies/standards and ending with future spectrum sharing games of mutually coordinating cognitive radios. The operation of 802.11 (Wi-Fi) and 802.16 (WiMAX) in unlicensed spectrum as an initial step towards technology-dependent horizontal spectrum sharing is described in Sections 5.1 and 5.2, respectively. Policies for spectrum sharing of dissimilar radio systems are described in Section 5.3. Section 5.4 gives a basic introduction to a spectrum policy framework, which is applied several times later in this book. Spectrum sharing games with solutions concepts derived from microeconomic game theory are described and evaluated on different levels in Section 5.5.

5.1 IEEE 802.11 in Unlicensed Spectrum

The protocols and transmission schemes of IEEE 802.11 Wireless Local Area Networks (WLANs) are truly one of the most remarkable of standardization achievements. An uncountable number of devices is today based on this standard, which started with a wireless extension for local area networks in 1997, and since then has been gradually improved and extended

Cognitive Radio and Dynamic Spectrum Access   L. Berlemann and S. Mangold
© 2009 John Wiley & Sons, Ltd

towards a very flexible, well-understood, technology. Originally, WLAN technology had been developed for enabling computer wireless network access. Today, WLANs are applied in consumer electronics, for instance mobile phones, gaming consoles and stand-alone Internet radios. Because 802.11 was built for radio systems in unlicensed spectrum, there is virtually no limitation on the use of 802.11: unlicensed spectrum is often harmonized throughout the world, which means that such radio systems can be used at any location and time. Wireless networks operating in unlicensed or open spectrum (see Section 2.1.1), are typically not designed for exchanging information between dissimilar radio networks. WLANs have to share, for instance, the unlicensed spectrum between each other and with other radio access technologies such as Bluetooth. Thus, the operation in unlicensed spectrum requires basic coexistence capabilities. The uncoordinated access to radio resources that are typically shared with other radio networks leads otherwise to increasingly problematic situations.

The basic coexistence capabilities of the 802.11 MAC are carrier sensing, collision avoidance, and contention-based channel access. They can be regarded as an initial step towards horizontal spectrum sharing of cognitive radio and are focus of this section. The 802.11 independent coexistence enablers of DFS and TPC are discussed in Section 3.2.1 in the general context of horizontal spectrum sharing.

5.1.1 Overview

802.11 has an hierarchical network architecture, as illustrated in Figure 5.1. Its basic element is the Basic Service Set (BSS), which is a group of stations controlled by a so-called

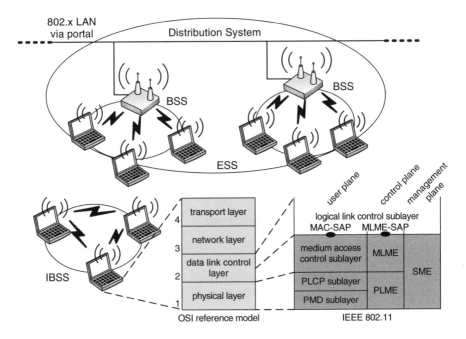

Figure 5.1 Basic architecture and reference model of 802.11

coordination function. The coordination function manages the access to the wireless medium and consists of two parts: (i) the Distributed Coordination Function (DCF) (described in detail in this section), which is used by all stations in the BSS, and (ii) the Point Coordination Function which is an optional extension for the support of QoS (not discussed here).

An Independent Basic Service Set (IBSS) is the simplest 802.11 network type. It is a network consisting of a minimum of two stations, where each station operates with exactly the same protocol. No station has priority over another, and the responsibility for coordinating medium access is distributed between all stations. An infrastructure-based BSS includes one station that has access to the wired network and which is therefore referred to the Access Point (AP). A BSS may also be part of a larger network, the so-called Extended Service Set (ESS). This ESS consists of one or more BSSs connected over a distribution system, for instance 802.3 Ethernet.

The 802.11 standard focuses on the lower layers one and two of the Open System Interconnection (OSI) reference model. This is indicated in the reference model of 802.11 as illustrated in the lower right part of Figure 5.1. This reference model divides the Data Link Control layer (i.e., OSI layer-2) into Logical Link Control (LLC) and Medium Access Control (MAC) sublayers. 802.11 defines Physical layer (PHY) transmission schemes (OSI layer-1), and the MAC protocol, but no LLC functionality. The LLC layer is independently specified for all 802 LANs standards, wireless or wired. Therefore, the management and control functions for addressing the characteristic implications of wireless communication systems need to be specified by 802.11 in the layers below. To consider, for example, radio range aspects, 802.11 includes functions for the management and maintenance of the radio network that exceed the usual MAC objectives.

5.1.2 Physical Layer

802.11 in its original form defines three different types of PHYs, namely 2.4-GHz Frequency Hopping Spread Spectrum (FHSS), Direct Sequence Spread Spectrum (DSSS) and Infrared (IR). There are five additional PHYs defined in the supplement standards 802.11a, 802.11b, 802.11g, 802.11n, and 802.11y, which are summarized in Table 5.1. Besides 802.11b, all of these PHYs are based on Orthogonal Frequency Division Multiplexing (OFDM). The 802.11 PHYs are developed for different operating frequency bands and channel bandwidths and differ essentially in their maximum throughput and transmission range. Details of these 802.11 PHYs can be found for instance in Chapter 5 of Walke *et al.* (2006).

Table 5.1 Overview of different physical layers of IEEE 802.11

IEEE 802.11	Frequency band (GHz)	Channel bandwidth (MHz)	Physical layer	Maximum data rate (Mb/s)	Transmission distance
802.11	2.4	20	FHSS/DSSS/IR	2	$\sim$20 m / $\sim$75 m
802.11a	5	20	OFDM	54	$\sim$30 m/ $\sim$100 m
802.11b	2.4	20	DSSS	11	$\sim$30 m/ $\sim$100 m
802.11g	2.4	20	OFDM	54	$\sim$30 m/ $\sim$100 m
802.11n	2.4, 5	20, 40	OFDM	248	$\sim$75 m/ $\sim$150 m
802.11y	3.7	5, 10, 20	OFDM	?	?/ $\sim$5000 m

5.1.3 Medium Access Control

In the following, the basic 802.11 MAC protocol, the DCF, is introduced. The DCF works as a listen-before-talk scheme, based on Carrier Sense Multiple Access (CSMA) and implementing a distributed medium access. 802.11 stations deliver MAC Service Data Units (MSDUs) of arbitrary lengths (up to 2304 byte), after detecting that there is no other transmission in progress on the radio channel.

5.1.3.1 Listen Before Talk

The 802.11 channel sensing function is called Clear Channel Assessment (CCA). It uses a single fixed power threshold, which is −82 dBm according to 802.11, but may be implementation dependent. If a station detects a signal with power larger than this threshold, the radio channel is assumed to be busy and thus unavailable for transmission. Otherwise, the radio channel is assumed to be idle. There are different variants of CCA. Either all signals are evaluated (ordinary noise detection), or only signals from other 802.11 stations are evaluated (noise plus preamble detection). See Figure 5.2 for an illustration of the CCA.

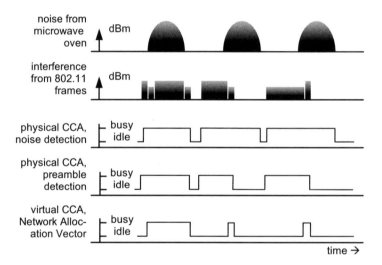

Figure 5.2   Carrier sensing in 802.11. A Clear Channel Assessment (CCA) is performed before each frame exchange based on noise detection, preamble detection or through channel reservation with the help of a network allocation vector

The Network Allocation Vector (NAV) is an addition to the physical sensing of the radio channel. It is referred to as virtual carrier sensing and in fact has the function of reserving the channel for some time. The NAV is a timer, which decreases irrespective of the status of the medium, which can be busy or idle.

The NAV is set when a frame is received that includes a duration field defining how long the following frame exchange may take. As long as the NAV is set or the CCA has sensed the radio channel as being busy, a station is not allowed to initiate transmissions. Thus, upon frame reception, the NAV can be eventually set for a duration that is longer than the transmission duration of this frame, and subsequent frame transmissions are protected.

5.1.3.2 Timing

According to the operation mode of the 802.11 DCF, each successful reception is acknowledged by the receiving station, as indicated on the left of Figure 5.3. The addressed station transmits an Acknowledgement (ACK) immediately after receiving a frame. The time between two MAC frames is called Interframe Space (IFS), and 802.11 defines four different IFSs. Short Interframe Space (SIFS), Point Coordination Function Interframe Space (PIFS) and Distributed Coordination Function Interframe Space (DIFS) are used under normal conditions and represent three different priority levels for medium access; the shorter the IFS, the higher is the priority in medium access. The fourth IFS, called Extended Interframe Space (EIFS), is used when a station detects an on-going transmission as being interfered with, assuming that there are some stations that cannot detect each other. A hidden station scenario is then assumed, and the station has to defer from channel access for a longer time.

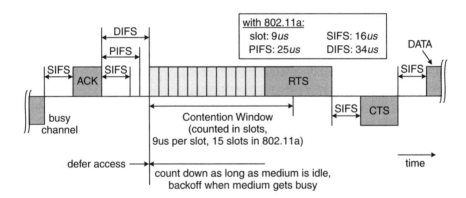

Figure 5.3 Contention-based channel access with random backoff of 802.11 (Walke *et al.*, 2006)

All interframe spaces are independent of the channel data rate. Due to the different characteristics of the different PHY specifications, the durations of the interframe spaces depend on the transmission scheme used. The relationships between the IFS and the duration aSlotTime (also referred to as slot time) are shown in Figure 5.3 for the example of 802.11a.

5.1.3.3 Collision Avoidance and Random Backoff

As part of the DCF, it may happen that more than one station will attempt to transmit at the same time. This is called a collision. In wireless communication, a transmitter cannot detect a collision at a receiver, while transmitting. To account for this, 802.11 is based on Carrier Sense Multiple Access/Collision Avoidance (CSMA/CA).

If two or more stations detect the channel as being idle at the same time, inevitably a collision occurs when these stations initiate a frame exchange at the same time. The 802.11

defines a CA mechanism to reduce the probability of such collisions. As part of CA, a station performs the so-called backoff procedure before starting a transmission. A station that has an MSDU to deliver has to keep sensing the channel for an additional random time duration after detecting the channel as being idle for the minimum duration DIFS, which is $34\,\mu s$ for 802.11a. Only if the channel remains idle for this additional random time duration is the station allowed to initiate its transmission. The duration of this random time is determined as a multiple of a slot duration (aSlotTime). Each station maintains a so-called Contention Window (CW), which is used to determine the number of slot times a station has to wait before transmission. Figure 5.3 shows an example in which after a successful frame exchange, i.e., after the ACK transmission, a station starts the next frame exchange (RTS frame followed by CTS frame), because the radio channel has been idle for a duration equal to DIFS and its following backoff slots. The contention window size increases when a transmission fails, i.e. when the transmitted data frame has not been acknowledged.

After an unsuccessful transmission, the next backoff is performed with a doubled size of contention window. This reduces the collision probability in case there are multiple stations attempting to access the channel. The stations that deferred from channel access during the channel busy period do not select a new random backoff time, but continue to count down the time of the deferred backoff already in progress after sensing the channel as being idle again. In this way, stations that deferred from medium access because their random backoff time was larger than the backoff time of other stations are given a higher priority when they resume the transmission attempt.

Figure 5.4 illustrates the increase of the contention window upon unsuccessful trans-missions. Note that a station cannot differentiate between collision and failed transmission due to errors on the wireless channels. A missed ACK frame will always be interpreted as collision.

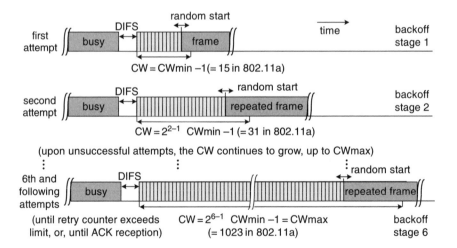

Figure 5.4 Exponential backoff of 802.11 for assisting collision avoidance. After unsuccessful frame exchanges, a random backoff is performed to determine the waiting time before the next transmission attempt. The random time is introduced by a contention window, which is doubled in size in case of collisions (Walke *et al.*, 2006)

5.1.3.4 Hidden Stations and Channel Reservation

In wireless communication systems that use carrier sensing, the so-called hidden station problem can occur, depending on the locations of the stations. This problem arises when a station is able to receive frames successfully from two different stations but the two stations cannot detect each other. When stations cannot detect each other, a station may sense the channel as idle even when other hidden stations are transmitting and may therefore initiate a transmission while the other station is already transmitting. This may result in a collisions and severely interfered frames at stations that can detect coinciding transmissions of hidden stations.

To reduce throughput reduction owing to hidden stations, 802.11 allows the optional use of a Request-to-Send/Clear-to-Send (RTS/CTS) mechanism. Before transmitting a frame, a station has the option of transmitting a short RTS frame, which must be followed by a CTS frame transmission by the receiving station. Between two consecutive frames in the sequence of RTS, CTS, data, and ACK, a SIFS, which is 16 μs for 802.11a, gives transceivers time to turn around. It is a decision made locally by the transmitting station, whether of not RTS/CTS is used. Upon receiving an RTS frame, the receiving station has to reply with a CTS frame. The RTS and CTS frames include information of how long it takes to transmit the next data frame, e.g., the first fragment, and the corresponding ACK frame. Hence, other stations close to the transmitting station and hidden stations close to the receiving station will not start any transmissions; their NAV timer is set. A hidden station close to the receiving station might not receive the RTS due to the large distance, but will in most cases receive the CTS frame. See Figure 5.5 for an example of the DCF using RTS/CTS. It is important to note that SIFS is shorter than DIFS, which always gives CTS and ACK the highest priority for access to the radio channel.

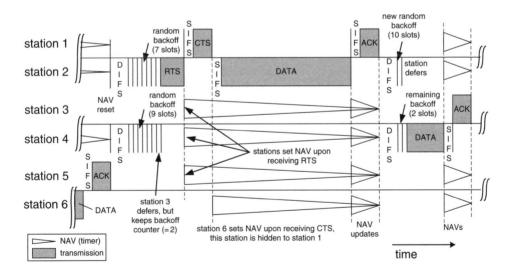

Figure 5.5 Timing diagram of frame exchanges and corresponding NAV settings in 802.11. Station 6 cannot detect the RTS frame of the transmitting station 2, but the CTS frame of station 1 (Walke *et al.*, 2006)

5.1.4 Learning from 802.11

We have concentrated in this section on aspects of the distributed medium access of 802.11 relevant to cognitive radio. The simplicity of contention-based medium access and the listen-before-talk principle has the promising potential of being applied for initial cognitive radio realizations. A detailed evaluation of the performance of 802.11 and its extensions for probabilistic QoS support can be found for instance in Chapter 5 of Walke *et al.* (2006).

WLANs of the 802.11 standard are able to coexist, i.e., operate at the same time and location without harmful interference in using their coexistence capabilities (see also Mangold *et al.*, 2001a). Nevertheless, the coexistence capabilities of today's WLANs are limited and fail when taking QoS support into account. Such coexistence scenarios, where distributed coordination for exclusive access is required, are not addressed in the IEEE 802.11 base standard (IEEE, 2007); see also Section 5.5.2. These scenarios are the motivation for our approach towards spectrum sharing games in Section 5.5.

5.2 IEEE 802.16 in Unlicensed Spectrum

Various scenarios might arise, where IEEE 802.16 (WiMAX) could have to share spectrum with already deployed and operating WLANs of IEEE 802.11 (Wi-Fi), such as in office or residential deployment scenarios. In particular, for supporting VoIP services, WiMAX may be more adequate than Wi-Fi.

There is no general regulatory objection against WiMAX operation in unlicensed spectrum apart from the usual regulatory restrictions of unlicensed operation in these frequency bands (see Chapter 2). The U-NII frequency band at 5 GHz is one example of spectrum that might be shared between 802.16 and 802.11a. Both radio access technologies are, for instance, candidates for implementing a Fixed Wireless Access (FWA) system in the unlicensed spectrum at 5.8 GHz for backhauling BSs. In such scenarios, the coexistence of 802.16 and 802.11 in shared radio spectrum is an acute problem. The frame-based medium access of 802.16 requires rigorous protection against interference from WLANs in order to operate properly when sharing spectrum.

As illustrated above, WLANs of the 802.11 standard are able to coexist. More complex strategies are required when QoS support is demanded: Successful, deterministic control of access to the radio resource is necessary for all coexisting wireless systems in order to guarantee QoS. The information exchange between spectrum sharing networks enables an interworking but is not required for coexistence. Approaches without information exchange based on the observation of spectrum utilization are discussed in the context of horizontal spectrum sharing in Section 3.1.3. With interworking, wireless networks are able to coordinate spectrum usage between each other. A central coordinating device that combines the central BS of 802.16 with the Hybrid Coordinator (HC) of 802.11e is proposed in Chapter 8 of Walke *et al.* (2006). This central device, called a Base Station Hybrid Coordinator (BSHC), requires operation in both protocol modes, 802.16 and 802.11(e), in order to realize interworking of WiMAX and Wi-Fi.

On the other hand, we here propose software upgrades limited to the MAC of the 802.16 BS. Through these upgrades, coexistence between 802.16 and 802.11 is enabled (without any data exchange) between both standards. Thereby, no 802.11 frame transmissions are required by an 802.16 system. We expect 802.16 systems to be available for laptops soon and can then provide wireless VoIP services that 802.11 cannot support satisfactorily well. As shown in this section, 802.16 is able to control access to a radio channel such that competing 802.11 systems only get access when permitted by an 802.16 BS. It is even possible for an 802.16 BS to push away any 802.11 system from a frequency channel at 5 GHz, if necessary.

5.2.1 Coexistence Scenario

Approaches for enabling the coexistence of a single 802.16 system with multiple 802.11a devices (APs and stations using the DCF for medium access) are introduced in the following. The basic idea is to prevent medium access of 802.11a before and during MAC frame transmission of 802.16. The proposed solutions target the avoidance of an idle medium with a duration equal or longer than DIFS.

The coexistence scenario considered in this section is illustrated in Figure 5.6, in which 802.16 BS and controlled 802.16 subscriber stations are depicted. They are operating at a frequency channel at 5 GHz that is shared with multiple 802.11a access points and stations. Details of the PHY of 802.11a and 802.16 are for example described in Walke *et al.* (2006). We assume that the 802.16 system has selected this frequency channel using DFS according to the regulatory restrictions given above. In the following, three risk points are identified that imply a danger for the interference free transmission of the 802.16 MAC frame. These dangers and their handling are illustrated in the timing diagram of Figure 5.7 and are marked with ①, ② and ③ respectively. It is worth mentioning again that all mechanisms proposed in this section have in common that no multimode device capable of operating according to 802.16 and 802.11 is required. A manipulation of the 802.11 devices' NAVs in order to prevent unwanted allocation attempts requires for example such a multimode device. This and similar concepts are discussed and have been evaluated (Walke *et al.*, 2006) under the umbrella of heterogeneous multihop networks.

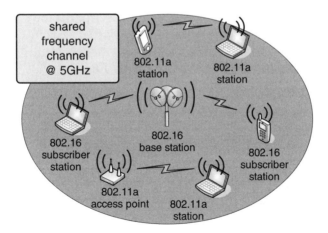

Figure 5.6 Coexistence scenario of 802.16 and 802.11a sharing the same frequency channel in the 5GHz band. Reproduced with permission from L. Berlemann, C. Hoymann, G. R. Hiertz, and B. H. Walke, 'Unlicensed Operation of IEEE 802.16: Coexistence with 802.11(a) in Shared Frequency Bands,' in Proc. of 17th IEEE Conference on Personal, Indoor and Mobile Radio Communications, PIMRC 2006, Helsinki, Finland, 11–14 September 2006. © 2006 IEEE

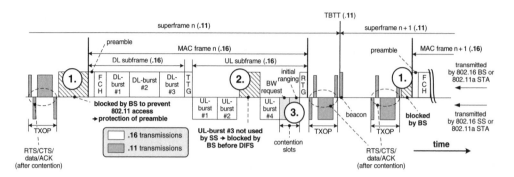

Figure 5.7 Timing diagram of an 802.16 MAC frame transmitted on a shared frequency channel. 802.16 BS protects the beginning of the MAC frame (①). Idle times longer than DIFS during the MAC frame are prevented – on the UL (②) and within contention (③). Additionally, the order of contention slots for initial ranging and bandwidth requests is reversed. Reproduced with permission from L. Berlemann, C. Hoymann, G. R. Hiertz, and B. H. Walke, 'Unlicensed Operation of IEEE 802.16: Coexistence with 802.11(a) in Shared Frequency Bands,' in Proc. of 17th IEEE Conference on Personal, Indoor and Mobile Radio Communications, PIMRC 2006, Helsinki, Finland, 11–14 September 2006. © 2006 IEEE

The MAC frame duration of 802.16 can be varied between 2.5 to 20 ms, while the beacon interval (often referred to as superframe) of 802.11 typically has a value of 100 ms. Thus, multiple 802.16 MAC frames are nested into one 802.11 superframe. We assume in the following, that the 802.16 system allocates only a fraction of the shared frequency channel for its own operation depending on its current traffic load. The time interval between two consecutive 802.16 MAC frames is accessed by the coexisting 802.11 devices in using the DCF, as also illustrated in Figure 5.7. We further assume that TTG and RTG are shorter than the DIFS duration interval of 802.11a.

5.2.2 Protecting the Beginning of 802.16 MAC Frame

Contrary to 802.11, 802.16 is not able to tolerate a delayed beginning of its MAC frame. Therefore, no 802.11a transmission is allowed to be ongoing when the 802.16 MAC frame is scheduled for beginning (with a preamble and the FCH). The BS therefore blocks the medium before the intended frame initialization in order to prevent an access from 802.11. The medium has to be blocked as soon as the time instance of the next 802.16 MAC frame is closer than the maximum duration of an 802.11 transmission. With the most robust PHY mode and the largest data packet size (2346 byte), the maximum duration of 802.11a transmission is approximately 2 ms. Thus, the blocked time interval has in the worst case a duration of 2 ms. An intended interfering with an ongoing 802.11 transmission that reaches into this blocking interval is not required and only a new allocation attempt after this transmission has to be prevented. Consequently, the effective duration of the blocked time interval may differ from one MAC frame to the next one, as also illustrated in Figure 5.7 (①). As the concrete beginning point of the time of the blocking is not under the control of the BS, the blocked time interval is difficult to use for 802.16 user data transmissions. The BS is not able to notify the 802.16 subscriber stations in the previous MAC frame about the exact beginning of the transmission used for blocking. Nevertheless, the BS may broadcast data on a best effort basis as a DL burst, although this data only has an unreliable chance of being scheduled for transmission.

In total, 20 % of the transmission time/capacity is wasted in the worst case for guaranteeing a timely beginning of the MAC frame. This can be seen as cost or effort for operating 802.16 in unlicensed frequency bands.

5.2.3 Protecting the 802.16 UL Subframe

The 802.16 system transmissions in the DL subframe are completely under the control of the BS and no 802.11 device has the opportunity to access the medium. The medium is never idle for a duration equal to or longer than DIFS. In case of the DL subframe this control is lost and a BS may not allocate an assigned UL-burst, so the medium may become idle again. In this case, the BS has to block the unallocated UL-burst with an own transmission before an idle time duration of DIFS, as depicted in Figure 5.7 (②). This is done to prevent a medium access of 802.11 and to protect the following UL-bursts of the on-going UL subframe.

5.2.4 Shifting the Contention Slots

The contention slots of 802.16 lead to a vulnerability of the 802.16 MAC frame in the face of potential 802.11a transmissions. These slots are used in the UL subframe for initial ranging and bandwidth requests by subscriber stations. The random access to these slots follows the slotted ALOHA principle. Unallocated 802.16 contention slots may lead to an idle medium with a duration equal or longer than DIFS. In this case, 802.11 devices might access the frequency channel and destroy the ongoing 802.16 MAC frame. The duration of contention slots in 802.16 depends on the used frequency band. Here, at 5 GHz with a bandwidth of 20 MHz, the contention slots used for bandwidth requests have a duration of 27.78 μs. The slots used for initial ranging are essentially longer and have a duration of 138.9 μs as summarized in Table 5.2. Consequently, the medium access of the 802.11a DCF after 34 μs implies a danger for the 802.16 MAC frame.

802.16 permits a rearrangement of the MAC frame structure. The contention slots are therefore scheduled after the end of the UL subframe in order to protect the UL bursts as depicted in Figure 5.7 (③). In this way, an interference of UL bursts due to an 802.11a medium access

Table 5.2 Basic OFDM parameters of 802.11a + 802.16 and resulting time values

Parameter	Value
802.11a + 802.16 – frequency band	At 5 GHz
802.11a + 802.16 – bandwidth	20 MHz
802.16 – OFDM symbol time	13.89 μs
802.16 – Contention slot duration for BW requests	Two OFDM Symbols = 27.78 μs
802.16 – contention slot duration for initial ranging	Ten OFDM Symbols = 138.9 μs
802.11a – DCF access after idle time of DIFS with a duration of	34 μs

in the contention phase of 802.16 is prevented. The contention slots for bandwidth requests are shorter than DIFS. At least the first two slots may be used for requests by subscriber stations without interference from 802.11a stations. The contention slots used for initial ranging are essentially longer than DIFS. Thus an access of 802.11a stations might only be prevented by the BS by blocking each unallocated contention slot if it is idle, similarly to the blocking of unused UL-bursts as described above. The order of the contention slots of the 802.16 base standard (IEEE, 2002) is therefore reversed here. First the slots for bandwidth requests are scheduled and thereafter the slots for initial ranging, as also illustrated in Figure 5.7 (③).

Ideally, associated subscriber stations do not need bandwidth requests in contention slots to change the capacity assigned to them by the BS. Usually this done by piggy backing these requests to scheduled UL transmissions in order to avoid the contention with other subscriber stations in the bandwidth request phase.

5.2.5 Quality of Service, Efficiency, and Fairness

This section introduces mechanisms for enabling the operation of IEEE 802.16 in spectrum shared with 802.11(a). Coexistence is enabled by partly blocking 802.11(a) out of the medium. This enables a guarantee of QoS in 802.16. A drawback to operating 802.16 in unlicensed frequency bands shared with 802.11 is identified, namely, in the worst case, about 20 % of the available transmission time is wasted, thus leading to a reduced efficiency of channel/spectrum usage.

The coexistence of multiple 802.16 BSs introduces an additional level of complexity, requiring intelligent algorithms for distributed coordination that imply intensive impacts on medium access control. This coexistence situation might for instance be mitigated in applying game theory as described in Section 5.5. Contrary, the solution proposed here requires only minimal modification of the medium access in the BS. The practical realization of the proposed solution is challenging but nevertheless feasible related to the required transceiver turnaround times in the 802.16 BS.

From the perspective of 802.11a, the proposed method can be regarded as unfair. Unfortunately, 802.16 requires such rigorous protection against interference in its MAC frame from other communication systems because of its inability to coexist. A fundamental regulatory recommendation for operation in unlicensed frequencies shared in the time domain can be derived from our proposed solution, which is limitation of spectrum access for its duration and the usage of deterministic spectrum allocation patterns, allowing a distributed coordination on the basis of spectrum observation. The fulfillment of this recommendation would enable more reliable QoS support (Berlemann, 2006), mitigate the coexistence problem of multiple 802.16 systems, and would improve fairness when sharing spectrum with dissimilar systems.

5.3 Policies in Spectrum Usage*

Flexible and dynamic spectrum usage requires intelligent medium access, especially in the face of QoS support. In this context, policies are required to restrict the dynamic spectrum

---

* L. Berlemann, S. Mangold, and B. H. Walke. 'Policy-based Reasoning for Spectrum Sharing in Cognitive Radio Networks,' in Proc. of 1st IEEE International Symposium on New Frontiers in Dynamic Spectrum Access Networks, DySPAN2005, Baltimore MD, USA, 8–11 November 2005. © 2005 IEEE

usage of cognitive radios. A policy is a selection of facts specifying spectrum usage. These facts are interpreted through a reasoning instance, in this chapter referred to as a spectrum navigator. The spectrum navigator is able to consider a flexible amount of different policies realizing a policy-adaptive cognitive radio. Policies, as regulatory rules for spectrum usage, form a framework for behavior of using spectrum. They are mandatory for operation and are enforced by regulation authorities.

Etiquette on the other hand adds fairness and efficiency to spectrum allocation. Etiquette is a multitude of rules that may be voluntarily applied and can be either part of standards or imposed by regulation. Spectrum etiquette is today already being discussed for existing unlicensed bands by various regulatory bodies and standardization groups.

5.3.1 Policy Framework

Policy enabled spectrum usage is one of the key features of cognitive radios. The decision taking and learning of a cognitive radio is not limited to policies but has to take many additional factors into account, like radio capabilities and the environment (outside world), such as the consumer. This imposes the need for a formal description framework. Initial steps towards a description language for cognitive radios have been introduced (Mitola and Maguire, 1999, 2000) as an ontology of radio knowledge as defined by Radio Knowledge Representation Language (RKRL). An initial step towards a policy frame work is the DARPA XG policy language (DARPA, 2004a) which includes an Extendable Markup Language (XML)-based policy description language. The SLS from Section 6.3 and strategies derived from game theory in Section 5.5 are described in the DARPA XG policy language at the end of the section. Additionally, the spectrum access of cognitive radios is specified in the DARPA XG policy language.

Policies have their origin in spectrum usage restrictions imposed by a regulating authority. Further policies may come from other policy makers to reflect, for example, preferences of the user or operators. The specification of algorithms for enabling spectrum sharing is another important aspect for using policies. The policies might have a limited validity that depends on multiple factors, such as, for example, the local time, the geographical location of the radio or the country where it is operating. A license holder may also impose policies for use its spectrum by a secondary radio system and might influence the access privileges to spectrum as well. Cognitive radios repeatedly look for updates of policies (say, once a day) that are relevant for their regulatory domain. The radios that are located in the regulatory domain for which new policies have been published, download the machine-understandable policies and update their local information bases. Alternatively, policies are made available through memory devices, such as flash cards, allowing cognitive radios that do not have access to servers to update their information bases. Thus, cognitive radios have to use policies in an adaptive way.

A well-defined policy framework is required to enable a cognitive radio capable of updating policies. This framework implies language constructs for specifying a policy, a machine-understandable representation of these policies, and a reasoning instance, here called spectrum navigator, which decides about spectrum usage as further outlined below. The policy conformance validation is responsible for downloading, updating and validating policies. The syntactical correctness of a policy that has been downloaded to a cognitive radio is verified. After conformity validation, the cognitive radio translates the policies to a machine-understandable language to enable computation through the spectrum navigator.

5.3.2 Spectrum Navigation

Cognitive radios have a flexible protocol stack and modem part that can both be dynamically adapted to the local communications environment. Additionally, a reconfiguration management is required to fulfill all reconfiguration-related functions. All functions concerning the opportunistic usage of frequency spectrum, i.e. realizing a cognitive medium access, are done by a spectrum navigator. This spectrum navigator is part of the reconfiguration plane (in the case of a completely reconfigurable protocol stack and modem part as, for example, considered in E^2R/E^3, see Section 4.1.3) or located in an 'open spectrum mode' (in case of a multimode capable radio of configurable modes such as, for example, under discussion in WINNER/WINNER+). The decision about how to allocate which spectrum is taken by the spectrum navigator on the basis of policies. The spectrum navigator identifies spectrum opportunities with the help of frequent measurements of spectrum usage provided by the protocol stack, for instance as under standardization in IEEE 802.11k as introduced above. There, means are developed for measurement, reporting, estimation and identification of the current spectrum usage in the ISM bands. Additionally, the QoS requirements of the supported applications are taken into account, together with preferences of the user, like transmission costs. The capabilities of a radio, like the frequency range that can be used for transmission, the available PHY modes, coding schemes, the number of transmission units, etc., determine which spectrum the navigator selects. The reasoning of the spectrum navigator results in specification of the current spectrum usage and corresponding configuration of the protocol stack, as depicted in Figure 5.8.

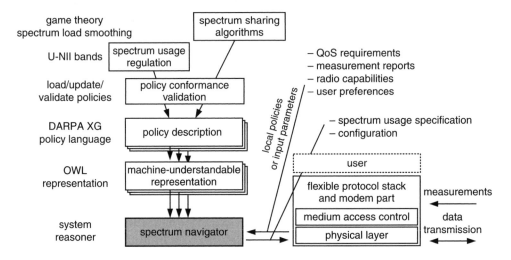

Figure 5.8 Flexible spectrum usage by a cognitive radio. A spectrum navigator takes multiple policies, spectrum usage measurements and additional restrictions into account when deciding about spectrum allocation. Reproduced with permission from L. Berlemann, S. Mangold, and B. H. Walke, 'Policy-based Reasoning for Spectrum Sharing in Cognitive Radio Networks,' in Proc. of 1st IEEE International Symposium on New Frontiers in Dynamic Spectrum Access Networks, DySPAN2005, Baltimore MD, USA, 8–11 November 2005. © 2005 IEEE

5.3.3 Reasoning Based Spectrum Navigation

This section is based on Mangold *et al.* (2005b). The variety of diverse understandings of what 'cognitive radio' refers to often leads to confusion. The many promises of what can be achieved if cognitive radios are employed lead, moreover, to high expectations about the cognitive radio approach. We therefore outline in this section the concept of reasoning as one of the core concepts for cognitive radio. This important aspect of cognitive radio is built on the DARPA XG vision (DARPA, 2004b). A cognitive radio is aware of its environment. 'Cognition' refers to an act of knowing, being aware, recognizing, judgment, and reasoning. Recent developments in the area of machine learning, the semantic web, and machine-understandable knowledge representation, allow the efficient implementation of a cognitive radio. It is realized in the form of a so-called reasoner, which is introduced above as spectrum navigator.

A reasoner makes the actual decisions as to how to share spectrum. A reasoner is a software process that uses a logical system to reach formal conclusions from logical assertions. It is able formally to prove or falsify a hypothesis, and is capable of inferring additional knowledge. The so-called first-order predicate logic is the simplest form of logical system considered to be useful for such a reasoner. As a simple example, a reasoner may be fed with the knowledge ('all cognitive radio devices are capable of operating at frequencies below 3.5 GHz'). A statement ('white space at 2.0 GHz') would enable this reasoner to infer ('spectrum usage permitted at 2.0 GHz').

Further information on traceability of decision-making and machine understandable radio semantics can be found in our vision of true cognitive radio later on in this book.

5.4 Policy Language

This section outlines the specification of policies with the help of an XML-based description language as an example of the DARPA XG Policy language (DARPA, 2004a).

As illustrated in Figure 5.9, a policy rule consists of three main elements. The first is the selector description that is used to filter policies to a specific environment. This is related,

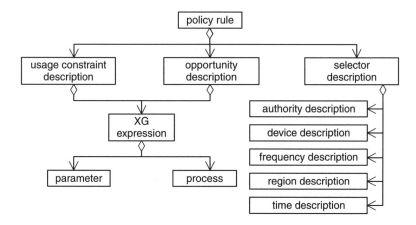

Figure 5.9 UML structure of policies in the initial DARPA XG Policy Language (DARPA, 2004a)

for example, to the policy issuing authority or the region where the policy is valid. Second is the opportunity description that specifies under which conditions spectrum is considered as unused. A specific received noise spectral density threshold is a simple example. Thirdly comes the usage constraint description that specifies the behavior of the cognitive radio when using a spectrum opportunity. All values that are contained in a policy, such as frequencies, levels/thresholds, or times, are described as parameters based on XML Schema Datatypes (XSD) (Biron *et al.*, 2004). Processes enable the execution of functions with input and output parameters. Measurements of the spectrum usage done by the protocol stack of the cognitive radio are an example of such a process.

Policy 5.1 illustrates the usage of the policy language with the example of the regulatory restrictions for using the Unlicensed National Information Infrastructure (U-NII) frequency band at 5 GHz in the US: an IEEE 802.11 Wireless Local Area Network (WLAN) device limits its transmission power to 40 mW when using the 5-GHz band. The lines have the following meanings:

Policy 5.1   Policies for using the U-NII Band at 5.15–5.25 GHz expressed in shorthand notation of the DARPA XG policy language. Reproduced with permission from L. Berlemann, S. Mangold, and B. H. Walke, 'Policy-based Reasoning for Spectrum Sharing in Cognitive Radio Networks,' in Proc. of 1st IEEE International Symposium on New Frontiers in Dynamic Spectrum Access Networks, DySPAN2005, Baltimore MD, USA, 8–11 November 2005. © 2005 IEEE

```
1    (SelDesc (id S1) (authDesc US-FCC) (freqDesc U-NII_US)
     (regnDesc US) (timeDesc Forever)
     (devcDesc 802.11device))

2    (DeviceDesc (id 802.11device) (deviceTyp WLAN_Class1)
     (deviceCap WLAN_Profile1))

3    (DeviceTyp (id WLAN_Class1))

4    (DeviceCap (id WLAN_Profile1)
     (hasPolicyDefinedParams MaxTransmitPower))

5    (FreqDesc U-NII_US
     (frequencyRanges U-NII_1 U-NII_2 UNII_3))

6    (FrequencyRange (id U-NII_1)
     (minValue 5.15) (maxValue 5.25) (unit GHz))

7    (Power (id TransmitLimit)
     (magnitude 40.0) (unit mW))

8    (Power (id MaxTransmitPower)
     (boundBy Device))

9    (UseDesc (id LimitTransmitPower)
     (xgx '(< = MaxTransmitPower TransmitLimit)'))

10   (PolicyRule (id P1) (selDesc S1)
     (deny FALSE) (oppDesc BandUnused)
     (useDesc LimitTransmitPower))
```

- Line 1 – The selector description S1 for a device named 802.11device. The issuing authority is the FCC (authDesc). The usage of the U-NII frequency band is described (freqDesc). The policies validity is limited to the US (regnDesc) and is not restricted to any period of time (timeDesc).
- Line 2 – The device 802.11device is described. It is of the type WLAN_Class1 and has the capabilities defined in WLAN_Profile1.
- Line 3 – A device type named WLAN_Class1. The meaning of the name is defined by a regulation authority.
- Line 4 – The capabilities of the device are defined as WLAN_Profile1. The device must understand and be able to provide the parameter MaxTransmitPower for computation in policies.
- Line 5 – U-NII Band consisting of three frequency bands.
- Line 6 – The U-NII Band at 5.15–5.25 GHz is described.
- Line 7 – A limit for the transmission power TransmitLimit is defined to 40 mW.
- Line 8 – MaxTransmitPower is declared and bound to a value provided by the protocol stack of the 802.11 device.
- Line 9 – Usage description of limiting MaxTransmitPower to TransmitLimit. xgx specifies an XG expression based on parameters to which the radio is able to provide values.
- Line 10 – Policy for using the U-NII Band at 5.15–5.25 GHz when it is regarded as opportunity described in BandUnused.

The opportunity description BandUnused and the frequency descriptions U-NII_2 and U-NII_3 are defined elsewhere.

Below, two approaches to distributed spectrum sharing are introduced and specified in DARPA XG policy language: The application of solution concepts derived from, on the one hand, game theory in Section 5.5.14, and the principle of SLS in Section 6.3.10 on the other.

Depending on the expressiveness and capabilities of a policy language and its framework, policies can also be used to specify simple spectrum access methods. A software-defined medium access control specified in the DARPA XG Policy Language is proposed later in this book. Initial steps towards the realization of a policy-based MAC will be discussed using the example of the Enhanced Distributed Channel Access (EDCA) of IEEE 802.11e. This channel access protocol is specified in a machine-understandable policy language instead of the lengthy textual description known from the standard. Such a machine-understandable description of the protocol enables cognitive radios to operate in distributed environments according to the 802.11(e) standard.

5.5 Spectrum Sharing Games*

Spectrum sharing in unlicensed bands is expected to become an increasingly significant research problem. With the proliferation of IEEE 802.11 wireless local area networks, and future cognitive radios using the unlicensed bands, spectral coexistence of dissimilar radio

---

* Reproduced with permission from: S. Mangold, L. Berlemann, and B. Walke. Equilibrium Analysis of Coexisting IEEE 802.11e Wireless LANs. In: 14th IEEE Conference on Personal, Indoor and Mobile Radio Communications, PIMRC 2003, Beijing, China, 7–10 September 2003, 321–325. © 2003 IEEE

systems has to be addressed in the future. In this section, we investigate how two independent wireless networks may share spectrum without direct coordination and information exchange. Solution concepts derived from the application of game theory in microeconomics are here applied to the competition for radio spectrum usage as a shared common resource. We use WLANs in unlicensed frequency bands as an example. The coexistence problem of spectrum sharing cognitive radios is approached using a stage-based, non-cooperative game (Friedman, 1971; Fudenberg and Tirole, 1998; Osborne and Rubinstein, 1994). The application of game models allows us to analyze the problem as a competition of players; within radio resource sharing games, payoff-maximizing players represent wireless networks. Decisions that players repeatedly have to make are about when, and how often, to attempt medium access. In this section's analysis, we use an established notation to describe multi stage strategies that determine player's decision making. Should radio networks cooperate, or should competing radio networks ignore the presence of other radio networks? What is the expected outcome in a repeated interaction? It is shown that traffic requirements imposed by services and applications determine the selected strategies that pursue cooperation or ignore other radio systems.

In this section, the terminology and notation of Neumann and Morgenstern (1953) and Osborne and Rubinstein (1994) are used. Many of the definitions can be found in Fudenberg and Tirole (1998), Debreu (1959) and Shubrik (1982). Widely accepted concise microeconomic standard text books on game theory are for example, Kreps (1990) and Mas-Colell *et al.* (1997). An electronic resource for educators and students interested in game theory can also be found here (gametheory.net, 2005).

Contrary to many other publications about the application of game theory in wireless communication, this section focuses on the decentralized coordination for distributed QoS support in unlicensed communications systems. The coexistence between broadband wireless access networks operating in unlicensed bands and WLANs (between WiMAX and Wi-Fi for example) can be another interesting application of this approach. The distributed coordination of reservations for spectrum allocation, for example in Wireless Personal Area Networks (WPANs) as part of the Distributed Reservation Protocol (DRP) (Hiertz *et al.*, 2003) in the MultiBand OFDM Alliance (MBOA) (Hiertz *et al.*, 2005a, b), is affected as well.

This chapter is outlined as follows. It begins with an introduction and detailed overview of the developed game model in Sections 5.5.3 and 5.5.4. The utility as an important part of the game model is introduced and defined in Section 5.5.5. An analytic model of for spectrum sharing as basis for the players' calculations of potential game outcomes is described in Section 5.5.6. In the game model of this book, Single Stage Games (SSGs) are the basis for interaction between spectrum sharing cognitive radios, here represented by players. The players are aware of the influence of their demanded allocations, i.e. spectrum usage, on the opponent ones. This enables an interaction, and thus the introduction of specific intended behaviors in Section 5.5.7. The interaction of the players within SSGs towards steady and thus predictable game outcomes is analyzed in detail under consideration of the microeconomic concepts of Nash Equilibria (NEs) and Pareto efficiency in Section 5.5.8. The game model of a Single Stage Game (SSG), including the behavior of a player, allows the introduction of an additional degree of interaction, then the dynamic interaction in repeated SSGs that form a multi stage Game (MSG) coordinated through strategies. Players estimate future expected outcomes of an MSG based on the discounted SSG payoffs as introduced in Section 5.5.10. The MSG outcomes depend on the players' strategy, which are differentiated in Section 5.5.11 into static, dynamic trigger and adaptive strategies. In the context of adaptive strategies, the establishment of cooperation through punishment, tolerance and forgiveness are discussed in

Section 5.5.11.4. The existence of Nash Equilibria (NEs), i.e., steady outcomes, within strategies on the basis of the payoff is analyzed in Section 5.5.12. The necessary step of abstraction in introducing the utility is reversed in Section 5.5.13 in evaluating the Quality-of-Service (QoS) outcomes of MSGs in dependency on the player's strategy.

5.5.1 Related Work

The following sections concentrate on the support of QoS by means of cooperation in decentralized wireless networks. Alternatively (Altman et al., 2002; Félegyházi et al., 2005; Srinivasan et al., 2003), game theory has been considered for cooperative relaying/routing in *ad hoc* networks. The delay minimization in slotted ALOHA (Ala Lokaho Oia'aha'a Ahonui) has been analyzed as a game problem (Altman et al., 2004) resulting in an insufficient system when applying game theory. Fundamentals of game theory and its application in resource management for modeling the interaction between service provider and customers are introduced by Das et al. (Das et al., 2004). Solution concepts from non-cooperative game theory are applied in this section, while cooperative game theory is, for example, considered by Yaiche et al. (Yaiche et al., 2000). The application of solution concepts derived from game theory for the analysis and realization of spectrum sharing has many facets. These are discussed in additional publications as summarized in Section 3.2.3.

The subsequent sections are based on a row of publications related to radio resource sharing games of coexisting wireless networks. Berlemann, Mangold and coworkers (Berlemann, 2002, 2003; Mangold, 2003; Mangold et al., 2003c) introduced the structure of a Single Stage Game (SSG), including a definition of the individual utility/payoff functions. Additionally, possible outcomes of SSGs are evaluated together with static strategies, i.e. strategies that do not adapt to changing conditions, in Multi Stage Games (MSGs). A simple Markov model to estimate throughputs for the radio resource sharing in SSG has been outlined (Mangold et al., 2003b). This model enables the players to select strategies for calculating potential game outcomes depending on the assumed strategy of the opponent. Dynamic strategies modeled as state machines and the possibilities of establishing cooperation through punishment are introduced by Berlemann and Mangold and coworkers (Berlemann et al., 2003a; Mangold et al., 2003a). The structure of an MSG, the preference of future payoffs (discounting) and the emerging Nash Equilibria (NEs) in MSGs of two players are described in detail by Berlemann et al. (2004a). Here, the ability of distributed QoS support is evaluated in focusing on the observed payoff in MSGs. The capability of punishment together with a comparison between game outcomes from aimed interaction and random play are illustrated elsewhere (Berlemann et al., 2004b) including a refinement of the utility/payoff functions. The outcomes of MSGs mutually depend on the strategies that are selected by all players. Berlemann and coworkers (Berlemann et al., 2005a) compared different dynamic strategies and discuss how they support QoS when players apply them. A tutorial-like description of this section's approach of using game theory for modeling the competition of WLANs sharing unlicensed frequency bands is given by Berlemann et al. (2005c). Strategies as policies described in the DARPA XG Policy Language (DARPA, 2004a) in the context of policy adaptive cognitive radios are given in the literature (Berlemann and Mangold, 2005; Berlemann et al., 2005e) and in Section 5.5.14.

5.5.2 802.11e Coexistence Scenario

A coexistence scenario of 802.11e WLANs according to the scenario illustrated in Figure 5.10, is considered in the following sections. Two 802.11e Hybrid Coordinators (HCs),

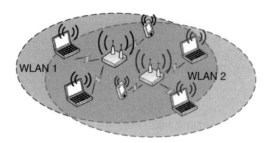

Figure 5.10 QoS support in spectrum sharing games of competing 802.11e WLANs for exclusive access to a single shared frequency channel. Hybrid coordinators of 802.11e WLANs are represented by players. Less prior EDCA traffic of 802.11e WLANs is represented by an additional player as random background traffic

each represented by a player, compete for exclusive access to a single, shared frequency channel. This is the classical coexistence problem known from 802.11/802.11e and is described in detail in Section 5.1. The exclusive right of the central coordinator to access the wireless medium is lost' a situation that is not considered in the standard. There, an isolated HC is assumed and potentially overlapping coverage areas of multiple HCs are neglected. A QoS guarantee can not be given. The results presented in Figure 5.11 illustrate the consequences of the coexistence on the observed QoS. The QoS of an isolated WLAN of one player, Figure 5.11(a), is compared with the QoS in a spectrum sharing scenario of two overlapping WLANs, Figure 5.11(b). In the first scenario a QoS guarantee is possible due to the means of 802.11e that are described in detail in Section 5.1. As the exclusiveness of spectrum access is lost in the second scenario, a QoS support is nearly impossible. This is indicated by the high transmission delays for both WLANs.

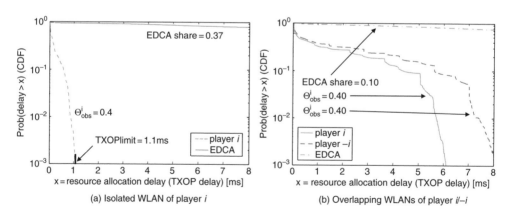

Figure 5.11 Coexistence problem of 802.11(e) WLANs. The QoS capabilities are evaluated in terms of observed throughput (Θ) and Transmission Opportunity (TXOP) delay. In (a), an isolated WLAN represented player i has exclusive access rights and a QoS guarantee is possible. In a scenario of overlapping WLANs (b), this exclusiveness is lost and the players' allocations interfere. Thus, a QoS guarantee is impossible

5.5.3 Game Overview

To facilitate the understanding of the terms used in this book, we illustrate the concepts with the help of the Universal Modeling Language (UML), see Figure 5.12. Each radio system is represented by a player, who competes with other players for control of a shared resource to support QoS. Such a player stands for all MAC entities of one coexisting wireless network. In our example case of coexisting IEEE 802.11e WLANs, a player represents (at least one) 802.11e HC. Although radio technologies for unlicensed bands share a common resource, they are typically not designed to arrange spectrum usage with different systems. We take this into account as we assume that players cannot establish communication with each other directly in order to coordinate their channel access. The QoS requirements imposed by services and applications define a multidimensional utility function. The game outcome, i.e. the utility under competition in the presence of other players, is referred to as payoff. The payoff is an abstract representation of the observed QoS. It is an important part of the stage-based game model and is discussed in detail in Section 5.5.4

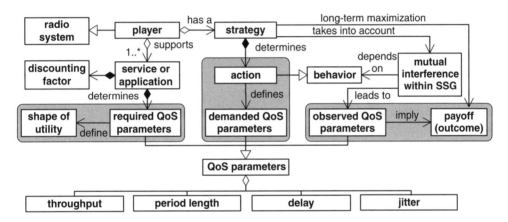

Figure 5.12 The fundamental terms of the game model in UML notation. A radio system is represented by a player. Each player 'has a strategy' for determining what action to select. An 'action specifies a behavior'. There are four QoS parameters: throughput, period length, delay and jitter. Reproduced with permission from L. Berlemann, B. Walke, and S. Mangold, 'Behavior Based Strategies in Radio Resource Sharing Games,' in Proc. of 15th IEEE International Symposium on Personal, Indoor and Mobile Radio Communications, PIMRC 2004, Barcelona, Spain, 5–8 September 2004. © 2004 IEEE

Players interact repeatedly by selecting their own behavior (a selection of MAC parameters). For the sake of simplicity, the behaviors of a player are limited here to cooperation and defection. After each stage of the game, the players estimate their opponent's behavior. The estimated behavior of the opponent has to be classified by taking its intention into account. This is discussed in detail in Section 5.5.6. The classification is necessary as there is no communication between the dissimilar radio systems, i.e. the players, which hinders direct negotiations. Nevertheless, players are aware of their influence on the opponent's utility, which enables interaction on the basis of punishment and cooperation, i.e., a handpicked allocation of the radio resource aiming at a specific intention.

The context of game theory for judging game outcomes is outlined on two levels. The existence of equilibria, i.e., steady outcomes of interaction, in SSGs depends on the players' behavior as outlined in Section 5.5.7. Strategies in a MSG, which is formed by repeated SSGs, are discussed on a second, higher-level in Section 5.5.11. The players decide about their strategy in discounting expected payoffs to calculate future outcomes of the MSG. The selected strategy of a player is decisive for the course of interaction. Thus, the ability to guarantee QoS depending on the players' strategies is evaluated in Section 5.5.13.

5.5.4 Single Stage Game for Frame Based Interaction

A SSG is introduced by an IEEE 802.11e superframe, i.e. beacon interval, and has a fixed time duration. Such a superframe is the time between two consecutive beacon frames that are used for system broadcasts. These beacons can be used to define such an interval, which is, in the following, arbitrarily set to a duration of 100 ms (the typical duration of an 802.11 superframe). Spectrum allocating players always allocate the shared channel exclusively with the help of the mechanisms of the HC, whose allocations are referred to as TXOPs. Note that each player represents all MAC entities of a single coexisting WLAN, and each network is assumed to include at least one HC. During the SSG duration, the demanded resource allocation times of all players determine which individual player can allocate a TXOP at what time.

An SSG consists of three phases, as illustrated in Figure 5.13, which form together one single stage: (I) The players decide about their action, which means they demand resource allocation times and durations. This is an instant of time at the beginning of an SSG, and hence does not consume any time at all. (II) The competitive medium access of the allocation

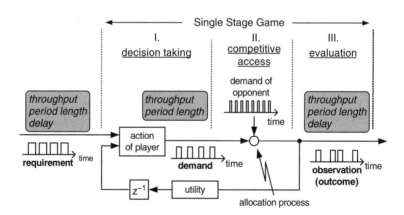

Figure 5.13 Single stage game for frame based repeated interaction. A single stage game consists of three phases: (I) The players decide about their action, (II) the competitive medium access, and (III) the players calculate game outcomes with the help of an individual utility function. The players might interact in repeated single stage games, which form a multi stage game. From L. Berlemann, S. Mangold, G. R. Hiertz, and B. Walke, 'Radio Resource Sharing Games: Enabling QoS Support in Unlicensed Bands,' IEEE Network, Special Issue on 'Wireless Local Area Networking: QoS Provision and Resource Management', vol. 19, no. 4, pp. 59–65, July/August 2005. © 2005 IEEE

process during the SSG occurs in the second phase. It may result in resource allocation delays and collisions of allocation attempts; resources may already be used by opponent players when an allocation attempt is demanded. Hence, the observed allocation points may differ from the demanded allocations, which are the reason for the difference of demand and observation. The second phase is the one that consumes the time of the SSG. (III) After the allocation process, players calculate the outcomes with the help of individually defined utility functions in the third phase of an SSG, again in an instant of time. The outcome (i.e. payoff) can be regarded as the difference between what was wanted and what is actually achieved. Spectrum allocations can be observed by all players, but the payoff as observed utility of a player, cannot be monitored by opponent players. The payoff is only individually known by each player. In the subsequent stage, demands and actual observations are taken into account when the players decide which action to select next. Observed outcomes of an SSG contribute to the game history over multiple stages, as will be explained in more detail in Section 5.5.9.

5.5.5 Quality-of-service as Utility

The QoS targets (values) that are considered in our game model are introduced in the following. These QoS targets are used to define a multidimensional payoff function as summarizing value of the observed QoS, which is the target of the players' efforts at optimization.

The game model of this subchapter comprises four abstract and normalized representations of QoS targets that are relevant in the context of distributed spectrum sharing:

- the throughput $\Theta \in [0, 1]$,
- the period length $\Delta \in [0, 0.1]$,
- the delay $\Xi \in [0, 0.1]$ and
- the jitter $\Psi \in [0, 0.1]$.

These QoS targets are normalized to values between 0 and 1. Period lengths, delays and the jitter are limited here to 0.1 to grant typical values for a stage length of 100 ms corresponding to 802.11 WLANs. The demanded throughput Θ and period length Δ determine a player's demanded allocations for an SSG as they are under the direct control of a player. They are selected at the beginning of each stage. This selection is the actual decision making, or, as it is called in the rest of this section, the *action* of a player. The period length, i.e., the interval between two successive TXOPs, aims to signal the players' tolerable delay; this period length, demanded by one player, can be observed by opponent players. Alternatively, the delay can not be observed as the originally requested allocation point of time is unknown to opponents. The period length is an important characteristic in the SSG. It allows players to estimate their opponent's intentions and to respond to their behavior. With the help of the period length, a player can signal his own intention, such as cooperation. Hence, this parameter allows the distributed establishment of cooperation, as is described below. The observed delay Ξ, i.e., as the difference between demanded and observed allocation points of time, is additionally considered in the utility/payoff functions to reflect the mutual interference resulting from the dynamics of the game. The jitter Ψ as fourth parameter, may be derived from the delay and is part of the classical QoS evaluation of the game outcomes from Section 5.5.13.

Observed delay and throughput can be significantly different to the demanded parameters, because they depend on the dynamics of the players' interactions during an SSG. This is reflected in the payoff function. The multidimensional payoff function represents the value of the observed QoS for a player. It is the unique objective of a player to optimize this value with respect to its QoS requirements defining the utility function. The payoff represents the observed and required QoS of a player and depends consequently on the above-introduced QoS targets. For example, a player may not be able to fulfill his QoS requirements (the payoff may be zero) although the radio resource is not frequently used, in times when a player is unable to allocate resources to required point of times.

5.5.5.1 Definition of Game Parameters

In the following equations the QoS targets from above are defined in using the parameters illustrated in Figure 5.14.

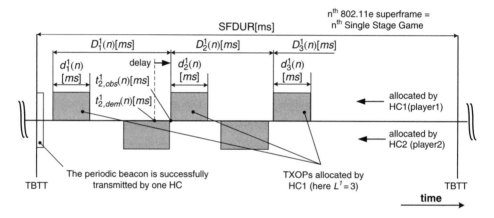

Figure 5.14 Allocation definitions within the nth superframe. It has a characteristic duration (SFDUR) of 100 ms. Each superframe forms a single stage game. Single stage games are repeatedly played and form a multi stage game

The throughput $\Theta^i(n)$ represents the share of capacity a player i demands in stage n of the game:

$$\Theta^i(n) = \frac{1}{SFDUR(n)} \sum_{l=1}^{L^i(n)} d_l^i(n) \tag{5.1}$$

where $L^i(n)$ is the number of allocated TXOPs per superframe n and $SFDUR(n)$ the duration of this superframe in ms, typically 100 ms. The parameter $d_l^i(n)$ describes the duration of the $TXOP$ l, $l = 1 \ldots L$, in ms, of player i in stage n. The variable $d_l^i(n)$ is illustrated in Figure 5.14.

The parameter $\Delta^i(n)$ specifies the maximum resource allocation period between subsequent resource allocations that player i observes or attempts to get in stage n. It is measured in ms and contains the delay between two TXOP transmissions due to the interrupted TXOP allocations, neglecting the effect that a transmission itself may fail and require retransmissions. $\Delta^i(n)$ is defined as:

$$\Delta^i(n) = \frac{1}{SFDUR(n)} \max \left[D^i_l(n) \right]_{l=1..L(n)-1}, \text{ with } D^i_l(n) \leq SFDUR(n), \quad (5.2)$$

where $D^i_l(n) = t^i_{l+1}(n) - t^i_l(n)$ is the time between the starting points of the two TXOPs l and $l+1$ of player i in superframe n, again measured in ms. See Figure 5.14 for an illustration. Note that $D^i_{L^i}(n) = t^i_1(n+1) - t^i_{L^i}(n)$ exceeds the superframe n.

The third QoS target is the delay $\Xi^i(n)$ of the allocated TXOPs for player i defined as

$$\Xi^i(n) = \frac{1}{SFDUR(n)} \max \left[\left| t^i_{l,obs}(n) - t^i_{l,dem}(n) \right| \right]. \quad (5.3)$$

The delay is defined as the difference between observed $t^i_{l,obs}(n)$ and demand $t^i_{l,dem}(n)$ allocation point of time of a TXOPs l of player i in superframe n. As this section concentrates on QoS guarantees, the focus of the game is rather on the maximum than on the mean values of Δ and Ξ.

The jitter Ψ as fourth QoS target may be directly derived from the delay:

$$d/dt \, \Xi^i(n). \quad (5.4)$$

For a detailed definition of these QoS targets see elsewhere (Mangold, 2003). In the following, the game model of Mangold (2003) and Berlemann (2002) is extended. This extension refers mainly to the utility/payoff function in order to take the dynamic interaction and the resulting consequences for the capabilities of supporting QoS into account.

The parameters d^i_l, D^i_l, L^i_l specify the allocation l of player i in detail. They are also depicted in Figure 5.14. To calculate these parameters out of a specific action, i.e., from the demanded QoS parameters, the following equations are needed:

$$L^i = \left\lceil \frac{1}{\Delta^i_{dem}} \right\rceil, \Delta^i_{dem} > 0, \quad D^i = SFDUR \cdot \Delta^i_{dem}, \quad d^i = SFDUR \cdot \Theta^i_{dem} \cdot \Delta^i_{dem}. \quad (5.5)$$

The players use these equations to determine their ideal allocation points of time and length for a single superframe without competitors. The competition influences the parameters L^i, D^i. The operator $\lceil \cdot \rceil$ rounds to the nearest integer value towards plus infinity.

The different roles in the context of the game model of these four QoS targets are shown in Table 5.3. Throughput, delay and jitter are considered in the classical QoS capability evaluation of this chapter's game model, done in Section 5.5.13. The utility/payoff function as defined in the next section takes the throughput, period length and observed delay into account. Again, this structuring is required to reflect the difference between QoS that can be observed by all players (here the observed throughput and period length) and QoS that is not observable in distributed environments (here the delay and jitter).

5.5.5.2 Action

The action of a player is defined based on the introduced QoS parameters, in the game called action $a^i(n)$ of player i in stage n of an MSG. Each player decides at the beginning of each

Table 5.3 Different roles of four QoS targets in the context of the game model

QoS target	Game parameter	Part of action	Represented in payoff	Part of QoS evaluation
Throughput	Θ	x	x	x
Period length	Δ	x	x	
Delay	Ξ		x	x
Jitter	Ψ			x

stage about its action. A simplified game of $N = 2$ players is assumed and the opponent is therefore called $-i$ in the following. As introduced above, an action of player i consists of the two demanded QoS parameters ($\Theta^i_{dem}, \Delta^i_{dem}$) and is defined as:

$$\begin{pmatrix} \Theta^i_{dem}(n) \\ \Delta^i_{dem}(n) \end{pmatrix} := \left(\begin{pmatrix} \Theta^i_{req} \\ \Delta^i_{req} \\ \Xi^i_{req} \end{pmatrix}, \begin{pmatrix} \widetilde{\Theta}^{-i}_{dem}(n-1) \\ \widetilde{\Delta}^{-i}_{dem}(n-1) \end{pmatrix}, H^n \right) \rightarrow a^i(n), a^i \in A^i \qquad (5.6)$$

An action depends on the players' QoS requirement ($\Theta^i_{req}, \Delta^i_{req}, \Xi^i_{req}$) and the expected action, i.e., demand, of the opponent ($\widetilde{\Theta}^{-i}_{dem}, \widetilde{\Delta}^{-i}_{dem}$). The superscript indicates that the action of the opponent is not known to the player but is estimated from the observed history of the game. Note that Ξ^i_{req} refers to the maximum tolerable delay and is introduced later in this section in the context of the utility function. In addition, the history H^n of own observed QoS parameters ($\Theta^i_{obs}, \Delta^i_{obs}, \Xi^i_{obs}$) up to the last stage $n-1$ is evaluated. Again, the QoS parameter Ξ is part of the utility function, introduced in the next section, but not of the action. A^i is the set of actions from which player i selects his action. This action set consists of an infinite number of alternative actions and is a single coherent subset of $\mathbb{R}^2$. The action set is a Euclidean space of two dimensions that is unempty and infinite but limited to QoS parameter specific intervals that are independent from the frame duration:

$$A^i = \begin{pmatrix} \Theta^i_{dem} \in [0, 1] \\ \Delta^i_{dem} \in [0, 0.1] \end{pmatrix} \qquad (5.7)$$

5.5.5.3 Definition of Utility Function

In general, a utility function is an ordinal quantity for the satisfaction of an individual entity. The players of our game model use a common resource, the radio spectrum. Radio spectrum is divided in the frequency domain into frequency channels. Their satisfaction depends on the individual requirements for spectrum usage.

Here, the utility $U^i \in \mathbb{R}^+_0$ defines what player i gains from a specific action $a^i(n)$ and is thus a set-based function over the action set A^i. The utility is an abstract representation of the supported QoS of a player and depends consequently on the above introduced QoS parameters. The definition of the utility function considers all characteristics of the QoS under consideration of the individual QoS requirements of a player, imposed by the supported applications. There are many different possible approaches to reflect QoS characteristics in a utility function. In this section, an approach based on rational and concave functions is chosen in order to simplify the analytical analysis. The following definitions of the utility functions are the results of in an intensive refinement process that covered the technical requirements of the utility function, the analytical usability and the expectations derived from game theory.

The utility of player i depends on three normalized utility terms $U_\Theta^i \in \mathbb{R}^+$, $U_\Delta^i \in \mathbb{R}^+$ and $U_\Xi^i \in \mathbb{R}^+$. They represent the observed share of capacity and points of time of resource allocation. The overall utility is given by:

$$U^i = U_\Theta^i \left(\Theta_{dem}^i, \Theta_{obs}^i, \Theta_{req}^i\right) \cdot U_\Delta^i \left(\Delta_{obs}^i, \Delta_{req}^i\right) \cdot U_\Xi^i \left(\Xi_{obs}^i, \Xi_{req}^i\right), U^i \in \mathbb{R}^+, \qquad (5.8)$$

where U^i is a nonnegative real number. All utility terms are dimensionless and have values between 0 and 1, hence $0 \le U^i \le 1$.

The utility function of the gained throughput U_Θ^i is defined as:

$$U_\Theta^i \left(\Theta_{dem}^i, \Theta_{obs}^i, \Theta_{req}^i\right) := \left(1 - \frac{1}{1 + u \cdot \left(\Theta_{obs}^i - \Theta_{req}^i + \Theta_{tolerance}^i\right)}\right) \left(1 + v \cdot \left(\Theta_{req}^i - \Theta_{dem}^i\right)\right)$$

$$(5.9)$$

if $\Theta_{obs}^i \ge \Theta_{req}^i - \Theta_{tolerance}^i$ and 0 otherwise. The parameters u and v in Equation (5.9) define the elasticity of the utility function and the tolerable deviation of the share of capacity is given as:

$$\Theta_{tolerance}^i = \frac{\sqrt{v^2 + u \cdot v} - v}{u \cdot v}, \quad u, v \in \mathbb{R}^+, \qquad u, v > 0.$$

Figure 5.15 illustrates the first utility term $U_\Theta^i(\Theta_{dem}^i, \Theta_{obs}^i, \Theta_{req}^i)$ depending on the two shaping parameters $u, v \in \mathbb{R}^+, u, v > 0$ for an exemplary $\Theta_{req}^i = 0.4$. The shaping parameters u and v have, in this section, values of 10 and 1 respectively. The resulting shape of the utility function is marked bold in the figure. As a function of the demand and of the observation, the utility function $U_\Theta^i(\Theta_{dem}^i, \Theta_{obs}^i, \Theta_{req}^i)$ extends the classical understanding of the utility as it is known from game theory (Fudenberg and Tirole, 1998). The dependency on the demand is added here because of the obvious necessity to restrict the players demanding a higher capacity than required. Access to the channel is limited by reducing the utility for high capacity demands, contrary to an otherwise gained higher utility from $\Theta_{dem}^i \to 1$. This reflects the fact that all players observe less utility in demanding high capacity at the same time. Demanding high capacity implies also poor results for the EDCA represented by an additional player as explained in Appendix B. The less prior EDCA is blocked out of the superframe.

To force the players not to allocate too much of the medium, the shaping factor v further reduces the observed utility for high values of Θ_{dem}^i to the benefit of the other players. The parameter u appends elasticity to the game model, and depending on this parameter the player may strictly need it or may be satisfied with less adequate observations.

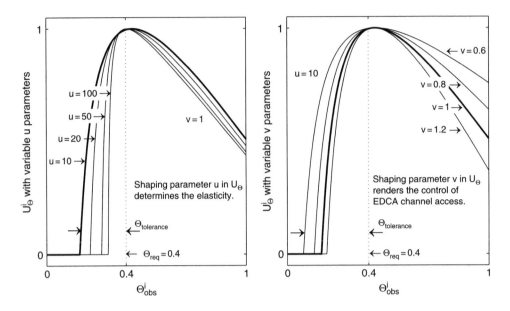

Figure 5.15 The utility term $U_\Theta^i(\Theta_{dem}^i, \Theta_{obs}^i, \Theta_{req}^i)$ for $\Theta_{req}^i = 0.4$. On the left side the shaping parameter u is varied and on the right v. The maximum utility is observed when the requirement is fulfilled

The second utility term U_Δ^i, which is related to the period length of resource allocations, depends on the same parameters u and v. Due to the smaller values of $\Delta \in [0..0.1]$, whereas $\Theta \in [0..1]$, the shaping parameters are multiplied by 10. U_Δ^i is defined as:

$$U_\Delta^i\left(\Delta_{obs}^i, \Delta_{req}^i\right) :=$$

$$\left(1 - \frac{1}{1 - 10 \cdot u \cdot \left(\Delta_{obs}^i - \Delta_{req}^i + \Delta_{tolerance}^i\right)}\right)\left(1 + 10 \cdot v \cdot \left(\Delta_{obs}^i - 2 \cdot \Delta_{req}^i + \Delta_{tolerance}^i\right)\right) \tag{5.10}$$

if $\Delta_{obs}^i \leq \Delta_{req}^i - \Delta_{tolerance}^i$ and 0 otherwise. The maximum length of allocation periods is related to the parameter $\Delta_{tolerance}^i$, which is defined as:

$$\Delta_{tolerance}^i = \frac{\sqrt{(10 \cdot v)^2 + 10 \cdot u \cdot 10 \cdot v} - 10 \cdot v}{10 \cdot u \cdot 10 \cdot v}, \quad u, v \in \mathbb{R}^+, \quad u, v > 0. \tag{5.11}$$

The absolute value of this parameter $\Delta_{tolerance}^i$ reflects the tolerated variation of the delay of resource allocations, i.e., TXOPs. Figure 5.16 depicts the second utility term $U_\Delta^i(\Delta_{obs}^i, \Delta_{req}^i)$ for varied shaping parameters u and v. The utility U_Δ^i can be compared to the mirrored U_Θ^i function. This reflects, in general, that a high Θ_{obs}^i and a low Δ_{obs}^i are preferable for a player. Real-time applications require constant allocation periods. Thus U_Δ^i is reduced for $\Delta_{obs}^i > \Delta_{req}^i$ and, in addition to this, it is not useful to have very short allocations, i.e., $\Delta_{obs}^i \ll \Delta_{req}^i$. The utility reaches its maximum value for $\Delta_{obs}^i = \Delta_{req}^i$ and decreases for small periods of resource allocation. The parameter v in $U_\Delta^i(\Delta_{obs}^i, \Delta_{req}^i)$ has consequently another intention as in $U_\Theta^i(\Theta_{dem}^i, \Theta_{obs}^i, \Theta_{req}^i)$, although it is used in the same way.

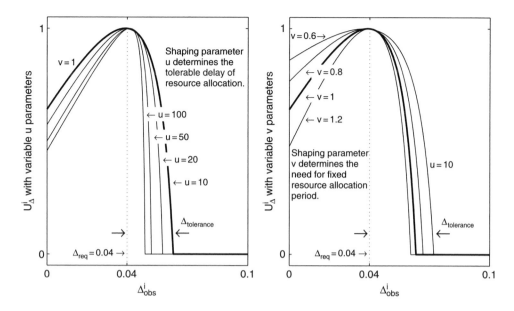

Figure 5.16 The utility term $U_\Delta^i(\Delta_{obs}^i, \Delta_{req}^i)$ for $\Delta_{req}^i = 0.04$. On the left side the shaping parameter u is varied and on the right v. The maximum utility is observed when the requirement is fulfilled

The third utility term is U_Ξ^i, which is related to the delay between the demanded and observed allocation points of time. U_Ξ^i depends on the convexity factor $w \in \mathbb{R}^+$ and is defined as:

$$U_\Xi^i\left(\Xi_{obs}^i, \Xi_{req}^i\right) := 1 - \frac{1 - e^{-w \cdot \left(1 + \frac{\Xi_{obs}^i - \Xi_{req}^i}{\Xi_{req}^i}\right)}}{1 - e^{-w}} \tag{5.12}$$

if $\Xi_{obs}^i \leq \Xi_{req}^i$ and 0 otherwise. Ξ_{req}^i reflects the maximum tolerable delay and is imposed by the supported applications of a player. Figure 5.17 illustrates the utility function $U_\Xi^i(\Xi_{obs}^i, \Xi_{req}^i)$ for a varied convexity factor w. For $w = 0$ the utility function is a straight line. In this subchapter, w has a constant value of 2 as indicated by the bold curve in Figure 5.17. The convexity of this utility function increases for increasing w. In this way the convexity factor w reflects the restriction on fixed allocation times or the permission of delays.

The shaping parameters u, v and w are used by all players through the complete game. All players have the same shaping parameters and these parameters are thus known to all players. In Equations (5.9), (5.10) and (5.12), the demand and observation of the player may change from SSG to SSG while the requirements remain constant for the complete MSG. This is assumed for evaluating steady game outcomes in the following sections.

For a better understanding of the utility function, Figure 5.18(a) depicts the observed utility of a player who exclusively utilizes the radio resource, depending on its demanded QoS (throughput and period length). The ideal case that no opponent player is present is assumed. Consequently, player i is observing its demand and its allocations are not delayed,

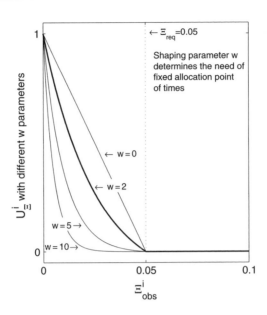

Figure 5.17 The utility term $U_\Xi^i(\Xi_{obs}^i, \Xi_{req}^i)$ for $\Xi_{req}^i = 0.05$. The convexity factor w is varied. For $w = 0$ the utility function forms a straight line. Reproduced with permission from L. Berlemann, G. R. Hiertz, B. Walke, and S. Mangold, Cooperation in Radio Resource Sharing Games of Adaptive Strategies. In: IEEE 60th Vehicular Technology Conference, VTC2004-Fall, Los Angeles CA, USA, 26–29 September 2004, 3004–3009. © 2004 IEEE

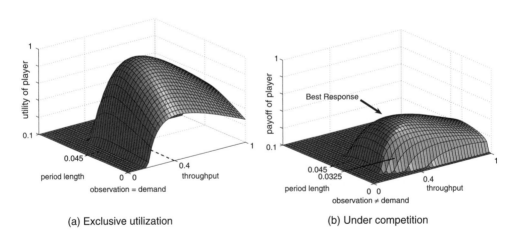

(a) Exclusive utilization   (b) Under competition

Figure 5.18 Equally scaled utility and payoff functions of a player depending on its demanded throughput and period length. In (a) no opponent is present while in (b) the unlicensed radio resource is under competition and shared with an opponent. Reproduced with permission from S. Mangold, L. Berlemann, and B. Walke, Equilibrium Analysis of Coexisting IEEE 802.11e Wireless LANs. In: 14th IEEE Conference on Personal, Indoor and Mobile Radio Communications, PIMRC 2003, Beijing, China, 7–10 September 2003, 321–325. © 2003 IEEE

i.e., $\Xi^i_{obs} = 0$. The utility function U^i is quasi-concave and has a unique maximum at the optimal fulfillment of the players QoS requirements. As introduced above, the player can demand a specific throughput, i.e. share of capacity, and period length that are together defined in Equation (5.7) as an action. The maximum of the utility function $U^i = 1$ is by definition given by the required QoS, here $(\Theta^i_{req}, \Delta^i_{req}) = (0.4, 0.045)$. Thus, a utility maximizing action (introduced as best response in the next section) can be defined as $\widehat{a}^i := (\widehat{\Theta}^i_{dem}, \widehat{\Delta}^i_{dem}) = (\Theta^i_{req}, \Delta^i_{req}) = (0.4, 0.045)$ in the case of there being no resource competition.

5.5.5.4 Payoff: Utility Under Competition

In a competition scenario of coexisting cognitive radios, the radio resource has to be shared. Under such competition, i.e. in the presence of another player, the players' allocations interfere. Thus, to evaluate the outcome of an SSG under competition, the opponent's action has to be taken into account as well. Therefore, the payoff V^i of player i in stage n is defined as utility under competition:

$$V^i(a^i, a^{-i}) \rightarrow U^i(a^i) \quad a^i \in A^i, a^{-i} \in A^{-i}. \tag{5.13}$$

The payoff describes what a player observes as utility by being a function of the actions, i.e. demands, of all participating players. In this way it is clearly separated between exclusive usage of the resource on the one hand, leading to $(\Theta^i_{obs}, \Delta^i_{obs}, \Xi^i_{obs}) = (\Theta^i_{dem}, \Delta^i_{dem}, 0)$ together with $U^i(a^i)$, and on the other hand, competition to access the resource resulting in $\Theta^i_{obs} < \Theta^i_{dem}, \Delta^i_{obs} > \Delta^i_{dem}, \Xi^i_{obs} > 0$ with $V^i(a^i, a^{-i})$. This leads consequently to $U^i(a^i) > V^i(a^i, a^{-i})$.

The payoff completes the SSG and highlights the dependency of player i's payoff V^i on the opponent's action a^{-i}. In addition, the observed QoS parameter Ξ^i_{obs} is now affected by the opponent's allocations and influences the payoff, too. Nevertheless it is not part of the player's action as defined in Equation (5.6).

Under competition, the players may observe less QoS than demanded. This leads to a decreased utility as depicted in Figure 5.18(b) on the same scale as Figure 5.18(a). The opponent has a QoS requirement of $(\Theta^i_{req}, \Delta^i_{req}, \Xi^i_{req}) = (0.4, 0.02, 0.02)$ leading to a fixed demand of $(\Theta^{-i}_{dem}, \Delta^{-i}_{dem}) = (0.4, 0.02)$ while the demanded throughput and period length of the player from above is varied. The observed delay, now inevitable because of the opponent's allocations of the shared radio resource, is considered in the utility function of Equation (5.8) as factor U^i_Ξ, but is not part of an action. U^i_Ξ influences the absolute value of U^i; with increasing observed delay Ξ^i_{obs}, which is still shorter than the required Ξ^i_{req} (maximal tolerated) delay, the observed utility decreases. Due to the competition, a player has to demand more restrictive QoS than is needed to satisfy its QoS requirement.

5.5.5.5 Best Response

Figure 5.18(b) introduces the best response action. When demanding resources a player has to consider the expected action of his opponent, which is a response to the opponent's action. Assuming the rational behavior of the opponent and using the history of interactions H^n, an opponent's action for the next stage can be estimated by the player (Mangold, 2003). The rational behavior of the opponent consequently results in an action that is again based on the opponent's assumption that all players act rationally. The question arises as to whether or not an action profile exists for which both players maximize their payoffs in the sense that neither player can improve his payoff by changing actions. This question is answered in Section 5.5.8.1 through the Nash equilibrium concept.

Through a variation of the assumed demanded Θ_{dem}^{-i} and Δ_{dem}^{-i} of player $-i$, player i can determine his best response on all possible actions of player $-i$ for an SSG. The payoff maximizing actions of player i to the opponent player's actions introduce a best response curve as illustrated in Figure 5.18(b). There, the best response is $(\widehat{\Theta}_{dem}^{i}, \widehat{\Delta}_{dem}^{i}) = (0.4, 0.0325)$. The players use an analytic model, introduced in Section 5.5.6, together with their belief about the opponent's expected action of the actual stage to calculate their potential payoffs and thus their best response. See Mangold (2003) for the detailed description of the technique for forming such a belief on the basis of the correlation of observed allocation patterns.

5.5.5.6 Information Structure

The information structure of the game model can be characterized as an asymmetrical distribution of information (Kreps, 1990; Mas-Colell *et al.*, 1997). The players are not aware of their opponents' QoS requirements and have no means of directly exchanging information. Thus they do not know the opponent's utility function exactly. Nevertheless, they are aware of the shape of the opponent's utility function, given by the shaping parameters u, v, and w. These parameters are assumed to be constant throughout a game. In addition, the players are aware of their action's qualitative influence on the opponent's utility. The players are able to determine more or less accurately the opponent's action for the last stage through observation, and can assume that this action will be identical to the actual stage's action. In MSGs, the players are able to form expectations about their opponent's action based on the observed game history H^n. The players are not aware of their opponent's behavior and consequently do not know the opponent's strategy. Nevertheless, the players can observe the opponent's influence on the own payoff together with their own strategy. This influence is quantitatively known to all players and so they are able to classify the opponent's intention and behavior as introduced in the next section.

*5.5.6 Analytic Game Model**

Before the actual play of an SSG, players must take their actions for a particular stage. This is performed by player i based on his own QoS requirements, which are given by Θ_{req}^{i}, Δ_{req}^{i}, taking into consideration the opponent player's demands $\widetilde{\Theta}_{dem}^{-i}$, $\widetilde{\Delta}_{dem}^{-i}$. The index $-i$ refers to the opponent of a player i. Note, that the superscript '~' indicates the fact that the demands of any opponent player $-i$ are not known to a player i, but are estimated from the history of the earlier stages of repeated SSGs.

In this section, a model for the game of two players that allows an analytical approximation of the expected observations as functions of the demands is presented. The approximation is

* S. Mangold, L. Berlemann, and B. Walke. Equilibrium Analysis of Coexisting IEEE 802.11e Wireless LANs. In: 14th IEEE Conference on Personal, Indoor and Mobile Radio Communications, PIMRC 2003, Beijing, China, 7–10 September 2003, 321–325. © 2003 IEEE

calculated by means of a Markov chain with five states. Note that in the rest of this chapter, the dependency of some game parameters on the game stage n is not indicated, since it is the SSG that being analyzed here.

5.5.6.1 Illustration and Transition Probabilities

In an SSG of two players, the calculation of the QoS observations is performed using the discrete-time Markov chain P illustrated in Figure 5.19 and defined by Equation (5.14). The longer the duration of an SSG and the higher the number of allocation attempts per stage, the more stationary the process becomes. A minimum of *ten* resource allocations per player is required, i.e. $\Delta_{dem}^{i,-i} < 0.1$; thus, a stationary state of the SSG is approximately achieved.

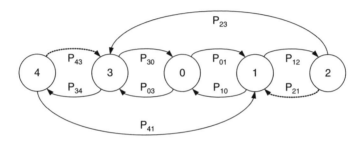

Figure 5.19   Discrete-time Markov chain P with five states to model a game of two players who attempt to allocate common resources. The default state in which the channel is idle or an EDCA frame exchange is ongoing is denoted as state 0. Reproduced with permission from S. Mangold, L. Berlemann, and B. Walke, Equilibrium Analysis of Coexisting IEEE 802.11e Wireless LANs. In: 14th IEEE Conference on Personal, Indoor and Mobile Radio Communications, PIMRC 2003, Beijing, China, 7–10 September 2003, 321–325. © 2003 IEEE

Further, it is assumed that none of the states is periodic. The aperiodic characteristic of P is a necessary condition for the game analysis, and cannot be assumed in general.

With $P_{03} = 1 - P_{01}$, $P_{10} = 1 - P_{12}$ and $P_{30} = 1 - P_{34}$, and by approximating $P_{21} \to 0$ and $P_{43} \to 0$, the corresponding transition probability matrix is denoted by:

$$
P = \begin{bmatrix}
0 & P_{01} & 0 & 1 - P_{01} & 0 \\
1 - P_{12} & 0 & P_{12} & 0 & 0 \\
0 & 0 & 0 & 1 & 0 \\
1 - P_{34} & 0 & 0 & 0 & P_{34} \\
0 & 1 & 0 & 0 & 0
\end{bmatrix}
\tag{5.14}
$$

The five states that the SSG process, which is modeled by P, can be in are introduced in Table 5.4. The resulting transition probabilities with $P_{ki} \geq 0, i, k = 0 \ldots 4$ are presented in Table 5.5.

Table 5.4 The five states of the Markov chain. Reproduced with permission from S. Mangold, L. Berlemann, and B. Walke, Equilibrium Analysis of Coexisting IEEE 802.11e Wireless LANs. In: 14th IEEE Conference on Personal, Indoor and Mobile Radio Communications, PIMRC 2003, Beijing, China, 7–10 September 2003, 321–325. © 2003 IEEE

State	Description
0	The channel is idle or allocated by low priority EDCA-TXOPs.
1	Player i successfully allocates resources with highest priority. Player $-i$ does not attempt to allocate resources.
2	Player i successfully allocates resources with highest priority, player $-i$ waits for the channel to become idle.
3	Player $-i$ successfully allocates resources with highest priority. This state is equivalent to state p_1 that models the same situation for the opponent player i.
4	Player $-i$ successfully allocates resources with highest priority, player i waits for the channel to get idle.

Table 5.5 The transition probabilities with $P_{ki} \geq 0$, i, $k = 0 \ldots 4$. Reproduced with permission from S. Mangold, L. Berlemann, and B. Walke, Equilibrium Analysis of Coexisting IEEE 802.11e Wireless LANs. In: 14th IEEE Conference on Personal, Indoor and Mobile Radio Communications, PIMRC 2003, Beijing, China, 7–10 September 2003, 321–325. © 2003 IEEE

Trans. Prob.	Description
P_{01}	Probability that player i allocates resources while the channel is idle or allocated by low priority EDCA-TXOPs.
P_{03}	Probability that player $-i$ allocates resources while the channel is idle or allocated by low priority EDCA-TXOPs that allocate resources via contention. $P_{03} = 1 - P_{01}$.
P_{10}	Probability that player $-i$ does not attempt to allocate resources during an ongoing resource allocation of player i.
P_{12}	Probability that player $-i$ attempts allocating resources during an ongoing resource allocation of player i, $P_{12} = 1 - P_{10}$.
P_{21}	Probability that player $-i$ gives up its attempt to allocate resources before player i finishes its resource allocation.
P_{23}	Probability that player $-i$ allocates resources right after player i finished its resource allocation.
P_{30}	Probability that player i does not attempt to allocate resources during resource allocation of player $-i$.
P_{34}	Probability that player i does attempt to allocate resources during resource allocation of player $-i$, thus, $P_{34} = 1 - P_{30}$.
P_{41}	Probability that player i gives up its attempt to allocate resources before player $-i$ finishes its resource allocation.
P_{43}	Probability that player i allocates resources right after player $-i$ finished its resource allocation.

5.5.6.2 Solution of the Markov Chain P

The stationary distributions of the Markov chain P can be calculated as:

$$
\begin{aligned}
p_0 &= 1 - p_1 - p_2 - p_3 - p_4 \,, \\
p_1 &= \frac{1}{2} \cdot \frac{P_{34} + P_{01} \cdot (1 - P_{34})}{1 + P_{34} + P_{01} \cdot (P_{12} - P_{34})} \,, \\
p_2 &= P_{12} \cdot p_1, \\
p_3 &= \frac{1}{2} \cdot \frac{P_{12} + (1 - P_{01}) \cdot (1 - P_{12})}{1 + P_{12} + (1 - P_{01}) \cdot (P_{34} - P_{12})} \,, \\
p_4 &= P_{34} \cdot p_3 \,.
\end{aligned}
\tag{5.15}
$$

Here, we assume $P_{23} \rightarrow 1$ and $P_{41} \rightarrow 1$ to address the fact that players tolerate delays in their resource allocation attempts, which occur when the opponent player allocates resources. It is assumed that a player never gives up his attempt to allocate resources while waiting for the opponent player to finish resource allocation. This implies a simplification that a player does not attempt to allocate more than one resource during one single ongoing resource allocation by the opponent player.

5.5.6.3 Transition Probabilities Expressed with QoS Demands

In this section, the QoS demands are used to determine the transition probabilities of P. The transition probability that player i allocates resources while the channel is idle or allocated by low priority EDCA-TXOPs via contention is approximated as:

$$
P_{01} \simeq \frac{L^i}{L^i + L^{-i}}, \quad L^i, L^{-i} > 0
\tag{5.16}
$$

During an SSG, the more TXOPs, L^i, player i attempts to allocate compared with the number of all high priority TXOPs, $L^i + L^{-i}$, the higher the probability of resource allocation of this player i. With $P_{03} = 1 - P_{01}$, the probability of resource allocation of player $-i$ can be calculated similarly.

It is further approximated that the transition probability for player $-i$ (i) attempting to allocate resources during an ongoing allocation of player $i(-i)$, is given by:

$$
P_{12(34)} \simeq min\left(1, \frac{d^{i(-i)}}{D^{-i(i)} - d^{-i(i)}}\right).
\tag{5.17}
$$

These transition probabilities are declared in a piecemeal way. The probability P_{12} is either $d^i/(D^{-i} - d^{-i})$ or approximated to 1, as expressed by Equation (5.17). The probability that player i decides to attempt a resource allocation while player $-i$ is allocating resources, depends on the ratio between the duration of this allocation d^i and the duration of the time interval between two consecutive demanded resource allocations of player $-i$, i.e., $D^{-i} - d^{-i}$. In the case that the time interval between two consecutive demanded resource allocations of player $-i$, given by $D^{-i} - d^{-i}$, is smaller than the duration d^i of a resource allocation by player i, player $-i$ will attempt to allocate resources immediately after the ongoing resource allocation, with probability i. For the reverse situation P_{34} is equivalently defined.

With the QoS demands as given in Equation (5.5), the transition probabilities of P are

$$
\begin{aligned}
P_{01} &= \frac{\Delta_{dem}^{-i}}{\Delta_{dem}^{-i} + \Delta_{dem}^{i}}, \quad \Delta_{dem}^{i,-i} > 0, \\
P_{12(34)} &= min\left(1, \frac{\Delta_{dem}^{i(-i)}}{\Delta_{dem}^{-i(i)}} \cdot \frac{\Theta_{dem}^{i(-i)}}{1 - \Theta_{dem}^{-i(i)}}\right), \quad \Delta_{dem}^{-i(i)} > 0, \Theta_{dem}^{-i(i)} < 1,
\end{aligned}
\tag{5.18}
$$

with $0 \leq P_{01}, P_{12}, P_{34} \leq 1$.

5.5.6.4 Average State Durations Expressed with QoS Demands

The average state durations T_0, T_1, T_2, T_3, T_4 are further required to calculate the QoS observations from the stationary distributions of P. The average duration of the model P being in the idle state, T_0, is approximated to:

$$
\begin{aligned}
T_0 &\simeq min\left(D^{-i} - d^{-i}, D^i - d^i\right) = \\
&SFDUR \cdot min\left(\Delta_{dem}^{i} \cdot \left(1 - \Theta_{dem}^{i}\right), \; \Delta_{dem}^{-i} \cdot \left(1 - \Theta_{dem}^{-i}\right)\right),
\end{aligned}
\tag{5.19}
$$

with the help of the QoS demands from Equation (5.5). This is understood as follows. If both players attempt to allocate resources periodically, the idle times between the resource allocations of a player i is denoted as $D^i - d^i$. In general, the player that requires shorter periods determines the average T_0 of the SSG. This is represented by the first part of Equation (5.19). The value of T_0 can be simplified to $T_0 \rightarrow 0$ for situations where the overall throughput demands of all involved players are relatively high, i.e., $\sum_i \Theta_{dem}^i \rightarrow 1$. In this case, it is very probable that the contention-based channel access through EDCA cannot allocate any resources, due to its low priority in medium access. Therefore, if $\sum_i \Theta_{dem}^i \rightarrow 1$, resources are nearly always allocated by one of the two players; the channel is busy most of the time.

The mean state duration is given by

$$
T_{Mean} \simeq p_0 T_0 + p_1 \cdot d^i + p_3 \cdot d^{-i}
\tag{5.20}
$$

because the duration of the process P being in state p_1 is determined by the duration of a resource allocation of player i, d^1, if the opponent player $-i$ does not decide to attempt resources during this allocation. In addition, if the opponent player decides to attempt a resource allocation during this allocation, the process changes to state p_2. The duration of the process P consecutively being in the states p_1 and p_2 is again determined by the duration of a resource allocation of player i, d^i. Therefore, it can be approximated that:

$$
p_1 T_1 + p_2 T_2 \approx p_1 \cdot d^i \quad and \quad p_3 T_3 + p_4 T_4 \approx p_3 \cdot d^{-i}.
\tag{5.21}
$$

The mean state duration T_{Mean} can now be expressed by using the QoS demands of Equation (5.5) as:

$$T_{Mean} \simeq SFDUR \cdot \left(\begin{array}{c} p_0 \cdot min \left(\Delta^i_{dem} \cdot \left(1 - \Theta^i_{dem}\right), \ \Delta^{-i}_{dem} \cdot \left(1 - \Theta^{-i}_{dem}\right)\right) \\ + p_1 \cdot \Theta^i_{dem} \cdot \Delta^i_{dem} + p_3 \cdot \Theta^{-i}_{dem} \cdot \Delta^{-i}_{dem} \end{array} \right) \tag{5.22}$$

where p_0, p_1, and p_3 are given through the solution of P, see Section 5.5.6.2. With this definition of the mean state duration T_{Mean}, the observed throughputs of the players are given by:

$$\Theta^{i(-i)}_{obs} = SFDUR \cdot \Delta^{i(-i)}_{dem} \cdot \Theta^{i(-i)}_{dem} \cdot \frac{p_{1(3)}}{T_{Mean}}. \tag{5.23}$$

Assuming a high offered traffic $\sum_i \Theta^i_{dem} \rightarrow 1$, with $P_{12} \rightarrow 1, P_{34} \rightarrow 1$ the throughput observation of player i is calculated as:

$$\Theta^i_{obs} = \frac{\Theta^i_{dem} \cdot \Delta^i_{dem}}{\Theta^i_{dem} \cdot \Delta^i_{dem} + \Theta^{-i}_{dem} \cdot \Delta^{-i}_{dem}}. \tag{5.24}$$

The maximum resource allocation period a player may observe due to delayed allocations during an SSG is calculated as:

$$\Delta^i_{obs} = \underbrace{\Delta^i_{dem}}_{\text{demanded allocation interval}} + \underbrace{\Delta^{-i}_{dem} \cdot \Theta^{-i}_{dem} + TXOPlimit}_{\text{unwanted increase of allocation interval (delay)}} \tag{5.25}$$

where the unwanted maximum increase of resource allocation intervals is dependent on the demand of the opponent player as well as the maximum duration of the EDCA-TXOPs. The latter is defined by the *TXOPlimit*. This *TXOPlimit* is neglected in the following as it was defined to be relatively small compared with the typical duration of resource allocations of the two players (*TXOPlimit* $\ll \Delta^i_{dem} \cdot \Theta^i_{dem}$ with $i \in N = \{1, 2\}$), and because of the lower priority in medium access through EDCA. Thus, the maximum observed resource allocation period is given by

$$\Delta^i_{obs} = \Delta^i_{dem} + \Delta^{-i}_{dem} \cdot \Theta^{-i}_{dem}. \tag{5.26}$$

The expected throughput observations $\Theta^{i,-i}_{obs}$ can be approximated by Equation (5.24), and for the observed allocation periods $\Delta^{i,-i}_{obs}$ an upper bound is given by Equation (5.26). In summary, with:

$$\Theta^i_{obs} \leq \Theta^i_{dem}, \quad \Delta^i_{obs} \geq \Delta^i_{dem}, \quad i \in N = \{1, 2\} \tag{5.27}$$

the model P results in the following analytical approximation for the observation of an SSG:

$$\begin{aligned} P := & \left(\left(\begin{array}{c} \Theta^i_{dem} \\ \Delta^i_{dem} \end{array} \right), \left(\begin{array}{c} \Theta^{-i}_{dem} \\ \Delta^{-i}_{dem} \end{array} \right) \right) \\ \rightarrow & \left(\begin{array}{c} \Theta^i_{obs} = min \left(\Theta^i_{dem}, \frac{\Theta^i_{dem} \cdot \Delta^i_{dem}}{\Theta^i_{dem} \cdot \Delta^i_{dem} + \Theta^{-i}_{dem} \cdot \Delta^{-i}_{dem}} \right) \\ \Delta^i_{obs} = \Delta^i_{dem} + \Delta^{-i}_{dem} \cdot \Theta^{-i}_{dem} \end{array} \right). \end{aligned} \tag{5.28}$$

5.5.6.5 Results of the Single Stage Game

In this section, a comparison of the model with simulation results is presented to assess how accurate the Markov model P represents the outcome of the SSG. In simulation, EDCA background traffic of 1 Mbit/s, with a TXOPlimit of $100\,\mu s$, was assumed. The analytical approximation does not capture the EDCA specifically. With SFDUR $= 100\,ms$, the maximum duration of the EDCA-TXOPs, defined by the TXOPlimit is smaller than the minimum duration of the resource allocations by the players. Hence, only minor influences on the game outcomes result from the EDCA.

Three different scenarios have been selected to review all relevant configurations. First, results are compared for a scenario where player 1 demands a shorter resource allocation interval $\Delta_{dem}^{1} = 0.02$ than it is demanded by player 2, $\Delta_{dem}^{2} = 0.03$, that means that $\Delta_{dem}^{1} < \Delta_{dem}^{2}$. Figure 5.20(a) shows the resulting outcomes of an SSG for both players, calculated with the analytical model P, as well as simulated. The demand for share of capacity of player 1 is varied between $\Theta_{dem}^{1} = 0$ and $\Theta_{dem}^{1} = 0.9$. The left hand side of Figure 5.20 shows the observed shares of capacity $\Theta_{obs}^{1,2}$ over the varying Θ_{dem}^{1}, and the right hand side shows the observed resource allocation intervals $\Delta_{obs}^{1,2}$ over Θ_{dem}^{1}.

It can be seen that the observed share of capacity increases with increasing demand up to a certain saturation point, according to simulation and analytical approximation (solid lines in the upper figure). The observed share of capacity of player 2 remains constantly at its demanded level, as long as the channel is not heavily overloaded (dotted lines in the upper figure). With heavy overload ($\Theta_{dem}^{1} > 0.8$), the approximation fails to model the effect of repeated collisions, which in general results in a loss of capacity for the player that demands the longer resource allocations, here player 2.

The right hand figure in Figure 5.20(a) shows the observed resource allocation intervals $\Delta_{obs}^{1,2}$. It can be seen that the observed resource allocation interval of player 2 increases with the increasing demand for share of capacity by player 1, which is again indicated by simulation and approximation. Note that an upper limit for the maximum observed resource allocation interval is approximated, according to Equation (5.28). The simulation results show some variations in the delay, which are a result of correlated resource allocation times and unpredictable collisions.

Although demanding $\Delta_{dem}^{1} = 0.02$, player 1 observes a larger resource allocation interval as this is obviously determined by the player demanding the longer resource allocations, here player 2. Simulation and analytical approximation show the maximum observed resource allocation interval within one SSG.

Secondly, in Figure 5.20(b) results are compared for a scenario where players 1 and 2 demand the same resource allocation interval $\Delta_{dem}^{1} = \Delta_{dem}^{2} = 0.02$. The observed share of capacity shows clear similarities in simulation and analytical approximation (left hand figure in Figure 5.20(b)). However, it can be seen that the approximated observations of the maximum resource allocation intervals are rather satisfying for player 1, but too pessimistic for player 2 (right hand figure in Figure 5.20(b)). This is due to the limitation of an upper limit rather than an expected value being approximated.

Finally, results are compared for a scenario where player 1 demands a longer resource allocation interval $\Delta_{dem}^{1} = 0.02$ than is demanded by player 2, $\Delta_{dem}^{2} = 0.01$, which means that $\Delta_{dem}^{1} > \Delta_{dem}^{2}$. From Figure 5.20(c) it can be observed that in this case the simulation results and the analytical approximation are very close to each other in nearly all cases.

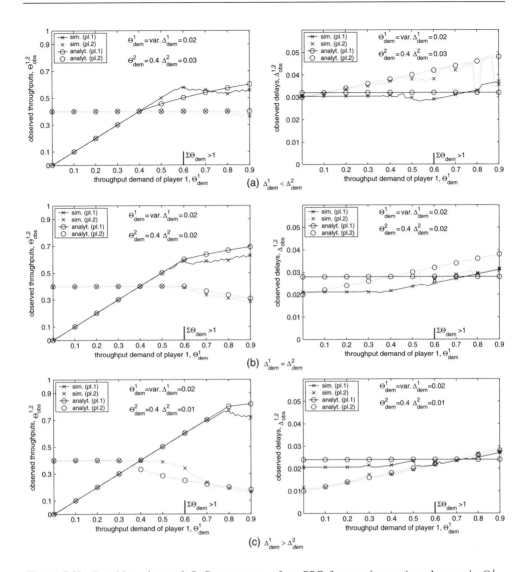

Figure 5.20 Resulting observed QoS parameters of an SSG for two interacting players via Θ^{l}_{dem}, calculated with P, and simulated. Left: observed share of capacity, right: observed resource allocation interval. Reproduced with permission from S. Mangold, L. Berlemann, and B. Walke, Equilibrium Analysis of Coexisting IEEE 802.11e Wireless LANs. In: 14th IEEE Conference on Personal, Indoor and Mobile Radio Communications, PIMRC 2003, Beijing, China, 7–10 September 2003, 321–325. © 2003 IEEE

The Markov model P was introduced for an analytical approximation of the outcome (observation) of an SSG. The main motivation was (i) to allow an analysis of the game on a purely analytical basis, and (ii) to allow a player to estimate possible outcomes of the game in advance while decision taking. Both goals are met. The model P is accurate enough to capture the statistical characteristics of the SSG. Whereas the model is simple enough to allow players to estimate the outcomes of an upcoming game in advance, this model can also be used for the detailed analysis of the SSG. In addition, the players are able to calculate the outcome of the SSG depending on their own and the opponent's expected actions. These results are considered when deciding what action to take and a further interaction of the players is enabled.

5.5.7 Behavior

The following section is based on a detailed analysis of potential SSG outcomes, performed by Mangold and by Berlemann (Mangold, 2003, Berlemann, 2002, 2003). The payoffs for both players depend on a single player's action. Therefore, SSGs resulting from various combinations of actions, i.e., ($\Theta_{dem}^{i,-i}, \Delta_{dem}^{i,-i}$), $\Theta_{dem}^{i,-i} \in [0, 1]$, $\Delta_{dem}^{i,-i} \in [0, 0.1]$, are evaluated in the following. It will be shown that the relative values of the demanded QoS parameters compared with the opponent's ones are decisive for a successful allocation of demanded resources.

5.5.7.1 Cooperation Through Predictable Behavior

In the absence of a centrally coordinating entity, a player can contribute to the establishment of cooperation by behaving predictably. Predictable resource allocations of a player during a SSG may allow other players to observe, understand and respond to his actions. Cooperation in the case of opponent cooperation is an example for this. For this reason, we refer to predictable behavior as a contribution to cooperation in the absence of a centrally coordinating entity (Mangold, 2003). The fixed periodicity of resource allocations by a player can be observed and predicted by other players. These other players may adapt their own resource allocations with the objective of mitigating mutual interference. Such behavior is cooperation as a response to an opponent's cooperative behavior.

A player's resource allocations reflect his individual QoS requirements. A decrease of the period lengths may however be considered as another important contribution to establishing cooperation (Mangold, 2003). While the number of equally distributed resource allocations per stage is increased, individual TXOPs (i.e., individual resource allocations) are shortened, which can reduce the observed delays of resource allocations of opponent players. In particular the CSMA/CA of 802.11 benefits from such an approach towards cooperation.

5.5.7.2 Definitions

As a result of the analysis of the SSG, behaviors are defined in this section. Contrary to the classical game theoretical understanding (Mas-Colell et al., 1997), behavior in the context of this section means a categorization of actions, not of strategies. A potential change of action consists of decreasing and increasing Θ_{dem}^i and Δ_{dem}^i corresponding to Equation (5.7). The definition of strategies is given in Section 5.5.11 through the dynamic change of behaviors within repeated SSGs. The behaviors are summarized in the action portfolio of Figure 5.21 later in this section.

Selfishness

The 'selfish' behavior can be characterized as:

$$\begin{pmatrix} \Theta^i_{dem} \\ \Delta^i_{dem} \end{pmatrix} = \begin{pmatrix} \gg \Theta^i_{req} \\ \Delta^i_{req} \end{pmatrix} \tag{5.29}$$

A player following selfish behavior solely focuses on his own payoff, independent of the opponent's action. In order to protect his own allocations, a player demands nearly the complete medium in order to block out the opponent. In leaving the demanded period length near its requirement, the selfish player's demanded allocation scheme reduces the medium access possibility for the opponent and at the same time maximizes his own payoff.

Best Response

A player i showing a of best response behavior selects an action corresponding to the highest expected payoff in the SSG as introduced in Section 5.5.5.5:

$$\left(\begin{pmatrix} \Theta^i_{req} \\ \Delta^i_{req} \\ \Xi^i_{req} \end{pmatrix}, \begin{pmatrix} \widetilde{\Theta}^{-i}_{dem} \\ \widetilde{\Delta}^{-i}_{dem} \end{pmatrix} \right) \Rightarrow \begin{pmatrix} \Theta^i_{dem} \\ \Delta^i_{dem} \end{pmatrix} \tag{5.30}$$

Best response behavior takes the expected opponent action for the current stage ($\widetilde{\Theta}^{-i}_{dem}, \widetilde{\Delta}^{-i}_{dem}$) into account when optimizing a player's own payoff, which again is a function of the individual QoS requirements ($\Theta^i_{req}, \Delta^i_{req}, \Xi^i_{req}$). Such behavior can be considered as rational. The SSGs of rational players are analyzed under the focus of the existence of NEs in Section 5.5.8. Depending on the players' requirement, the game outcome of the SSG, i.e. resulting payoffs, may not be Pareto efficient, see Section 5.5.8.3.

Punishment

Punishment can be implemented in different ways. All actions of a player that reduce the utility of the opponent can be considered as punishment, because the player is aware of its influence on the opponent. Demanding more restrictive QoS targets than needed, sending a busy tone or transmitting empty data packets are examples of punishing the opponent.

To analyze a potential interaction with the opponent, a player must be aware of its influence on the opponent's payoff, i.e. in which way his own action increases or decreases the opponent's payoff. Cooperation, as introduced below, leads to an increase of the payoff. To influence the game outcome to his own advantage, a player wants the opponent to cooperate too. The threat of potential punishment may lead to an establishment of game-wide cooperation, as discussed by Berlemann (2006).

The QoS requirements are most decisive for the ability to punish and the danger of being severely punished. They determine the minimum payoff a player may observe in case of a behavior following a payoff maximizing best response and punishment by the opponent. If the QoS requirements are very restrictive, the payoff is heavily reduced by a relative small deviation of observation from the requirement. Such a player can be considered as weak in defending his own allocations against the opponent. Contrary, a player with a robust requirement is immune to payoff reductions resulting from small observation deviations. A detailed analysis of the SSG outcomes, as performed by Berlemann and Mangold (Berlemann, 2002; Mangold, 2003), leads to the result below. Player i is stronger than his opponent $-i$ for:

$$\Theta^i_{req} > \Theta^{-i}_{req}, \Delta^i_{req} > \Delta^{-i}_{req}. \tag{5.31}$$

Directly related to the question of strength is the issue of punishment. A player with requirements leading to actions that imply strength in resource allocation is much more difficult to punish through defection than a weak one. The most successful punishment of the opponent is achieved by demanding the maximum values of the QoS parameters, by means of

$$\begin{pmatrix} \Theta^i_{dem} \\ \Delta^i_{dem} \end{pmatrix} = \begin{pmatrix} 1 \\ 0.1 \end{pmatrix}. \tag{5.32}$$

The punishment of the opponent affects also the player's payoff. Therefore, a less effective but nevertheless payoff-related punishment is introduced in connection with the definition of 'Defection' below.

Cooperation

A player i showing the behavior of 'cooperation' (C) attempts to gain higher payoffs than would result from a game of best response behaving players. The C behavior of player i allows the opponent player $-i$ to meet his requirements better although player i is not aware of the opponent's QoS requirements. In the case of a same-behaving opponent, all players benefit from cooperation. The cooperation is characterized as being beneficial for the opponent. At the same time, C leads to a vulnerable position for the cooperating player and often results in the highest payoffs in the case of game-wide cooperation. To define C behavior, analytical evaluations and simulations have been performed under consideration of these characteristics, leading to the results discussed by Berlemann (2002) and Mangold (2003).

This analysis of potential SSG outcomes leads to the following definition of cooperation: a cooperating player demands his required throughput and, independent of the requirement, a constant short period length of Δ_{min}, leading to:

$$\begin{pmatrix} \Theta^i_{dem} \\ \Delta^i_{dem} \end{pmatrix} = \begin{pmatrix} \Theta^i_{req} \\ \Delta_{min} \end{pmatrix}. \tag{5.33}$$

An analysis of the tradeoff between collision probability and short period lengths, i.e., high number of short allocations per stage, leads to the definition of $\Delta_{min} = 0.008$. This value results from a row of simulations that have indicated that it is a good compromise between high collision probability and less frequent idle times (Berlemann, 2002; Mangold, 2003). Through the common, requirement independent Δ_{min}, all players have nearly the same strength within an SSG. This parameter might for instance be introduced by a regulation authority as part of a spectrum etiquette to guarantee fairness in spectrum access.

Defection

The definition of the behavior of 'defection' (D) reflects two aspects. On the one hand, D can be an intended act of ending an established game-wide cooperation for the purpose of increasing own payoff. On the other hand, D can be the reaction to an opponent's deviation from game-wide cooperation with the aim of punishing the opponent in response. There are two options for defining a defective behavior: (i) 'punishment', or (ii) 'best response' behavior. The first possibility is very successful in reducing the opponent's payoff but might lead, on the other hand, to severe own payoff reduction. Alternatively, the 'best response,' as defection reflects the awareness of the players that a busy tone as defection is not suitable for providing interaction. Two players defecting with busy tones are not able to interact and support any QoS. Therefore, the best response, as payoff maximization under consideration of an existing opponent, is much more suitable for the game model. The player maximizes his own payoff while reducing the opponent's payoff. Thus the behavior of D corresponds in this section to the best response (above).

Defective behavior implies a negligence of future payoffs motivated, for example, by the players' support of applications on a 'best-effort' basis such as e-mail. Applications with restrictive QoS requirements, for example video conferencing services, give a reason for cooperative behavior.

5.5.7.3 Action Portfolio: Summary of Available Behaviors

The actions that are available to a player, i.e., all combinations of demanded QoS parameters Θ^i_{dem} and Δ^i_{dem}, are illustrated in the action portfolio in Figure 5.21. Exemplary allocation schemes are depicted in each corner to illustrate the dependency of the allocations on the demanded QoS parameters. The case of Figure 5.21, (a), on the left can be compared to leaving the game; (b), on the right, is an occupation of all resources for all time with a busy tone as introduced above as one option for realizing punishment.

An action in the area of selfish behavior leads to an aggressive allocation scheme. A selfish player allocates a high share of capacity, i.e., long resource allocations, which are not necessary for to its current QoS requirements. Hence, this behavior can be interpreted as blocking out the opponent completely when demanding a high share of capacity without regarding its own QoS requirements.

Cooperation behavior intends to gain highest payoffs in the case of game-wide cooperation. Initially, cooperation allows only the opponent player $-i$ to meet his requirements. Player i benefits from cooperation in the case of a cooperating opponent. Nevertheless, cooperation implies weakness against a non-cooperating opponent because the cooperating player can easily be blocked out.

5.5.7.4 Behavior as Basis for Interaction

From game history H^n the players form an individual table of expected payoffs on the basis of the common payoff table introduced in Table 5.6(a). This table defines the expected payoffs from an SSG of the two players depending on the classification of C or D. Player i's C/D behavior is in the left-hand column, player $-i$'s in the upper row.

The game model and the basic IEEE 802.11e access mechanisms, i.e. the EDCA, to a shared resource are evaluated with the help of the simulator YouShi, which is briefly described in Appendix B. To outline potential interactions on the basis of the player's behavior, Figure 5.22 illustrates a large set of SSGs simulated with YouShi. The observed payoffs in SSGs for both players, V^i_{SSG} for player i (Figure 5.22a) and V^{-i}_{SSG} of player $-i$

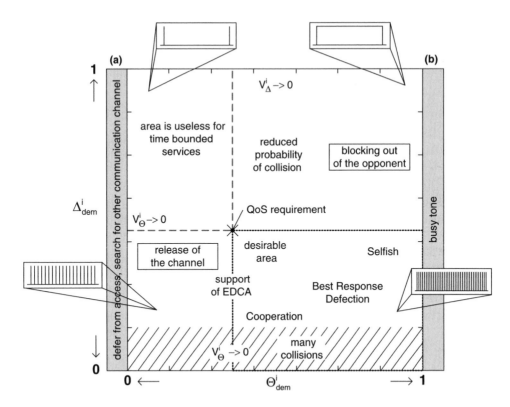

Figure 5.21 Portfolio of available actions. The corresponding payoffs and the resulting consequences on the opponents are depicted. Reproduced with permission from S. Mangold, L. Berlemann, and B. Walke, Equilibrium Analysis of Coexisting IEEE 802.11e Wireless LANs. In: 14th IEEE Conference on Personal, Indoor and Mobile Radio Communications, PIMRC 2003, Beijing, China, 7–10 September 2003, 321–325. © 2003 IEEE

(Figure 5.22b) are shown for different QoS requirements (Θ^i_{req}, Δ^i_{req}, Ξ^i_{req}) of player i on the abscissa as well as depending on the behavior of both players, denoted as (behavior of player i, behavior of player $-i$), on the ordinate. One required QoS parameter of player i is varied in each game scenario while the opponent's QoS requirements remain fixed at (Θ^{-i}_{req}, Δ^{-i}_{req}, Ξ^{-i}_{req}) = (0.4, 0.031, 0.02). The observed payoffs are averaged over the outcomes of 400 stages of two player SSGs. A third participating player, who is not evaluated here but who is part of the QoS evaluation of Section 5.5.13 represents the contention-based access of both Wireless Local Area Networks (WLANs) (in IEEE 802.11e referred to as EDCA) with an overall load of 5 Mb/s.

Two game scenarios are identified (dark gray marked) and transferred for combination to Tables 5.6(b) and (c), without limiting the generality of the game approach. The left four gray marked values in Figures 5.22(a) and (b) form the game scenario I in Table 5.6(b) with (Θ^i_{req}, Δ^i_{req}, Ξ^i_{req}) = (0.4, 0.05, 0.02), while the right gray marked values form game scenario II in Table 5.6(c) with (Θ^i_{req}, Δ^i_{req}, Ξ^i_{req}) = (0.1, 0.05, 0.02). An SSG-based analysis of these two scenarios shows that in Table 5.6(b) behavior '(D,C)' and Table 5.6(c)

Table 5.6 (a) Common payoff table. (b) Game scenario I: there is one NE (marked gray) as also illustrated in Figure 5.23, (D, C), namely (0.71,0.35). Three outcomes are Pareto efficient, only (D, D) is not Pareto efficient. (c) Game scenario II – 'Prisoner's dilemma': There exists one NE (marked gray) as illustrated in Figure 5.24, (D, D), namely (0.38,0.59). All outcomes are Pareto efficient except for this NE (D, D). Reproduced with permission from L. Berlemann, G. R. Hiertz, B. Walke, and S. Mangold, Cooperation in Radio Resource Sharing Games of Adaptive Strategies. In: IEEE 60th Vehicular Technology Conference, VTC2004-Fall, Los Angeles CA, USA, 26–29 September 2004, 3004–3009. © 2004 IEEE

(a) Common payoff table

Pl. $i \downarrow$ Pl. $-i \rightarrow$	D	C
D	V_{DD}^{i}, V_{DD}^{-i}	V_{DC}^{i}, V_{DC}^{-i}
C	V_{CD}^{i}, V_{CD}^{-i}	V_{CC}^{i}, V_{CC}^{-i}

(b) Game scenario I

Pl. $i \downarrow$ Pl. $-i \rightarrow$	D	C
D	(0.32,0.04)	(0.71,0.35)
C	(0.24,0.77)	(0.44,0.70)

(c) Game scenario II - 'Prisoner's dilemma'

Pl. $i \downarrow$ Pl. $-i \rightarrow$	D	C
D	(0.38,0.59)	(0.69,0.44)
C	(0.27,0.78)	(0.44,0.70)

behavior '(D,D)' is the unique NE marked gray in the respective table. The NE is thus a stable operation point, where neither player can gain a higher payoff as introduced below in Section 5.5.8.1.

Within a game scenario the relative difference between the players' payoffs depending on the own and opponent's behavior is decisive for the course of interaction and the existence of cooperative game outcomes. From player i's point of view, the payoff difference of $V_{CC}^{i} - V_{CD}^{i}$ implies the strength of punishment after an own defection and return back to cooperation thereafter. The benefit from cooperation may be considered as $V_{CC}^{i} - V_{DD}^{i}$. The term $V_{CC}^{i} - V_{DC}^{i}$ reflects the temptation to defect. Additionally, the difference of $V_{DD}^{i} - V_{CD}^{i}$ can be regarded as the possibility of rescuing the own payoff after opponent defection. In game scenario I the opponent player $-i$ is not able to increase his payoff by switching from cooperation to defection, because of $V_{DD}^{-i} < V_{DC}^{-i}$, contrary to game scenario II. This is also depicted in Figure 5.22(b) and leads to the different NEs. For further clarification, it is referred to the bargaining domains depicted in Figures 5.23 and 5.24, which are discussed in Section 5.5.8.2.

The players are able to make conclusions from their own observed payoff in an SSG on the opponent's behavior by consideration of their own behavior. From player i's point of view classification is realized through the statement of:

$$V_{SSG}^{i}(C,D) < V_{SSG}^{i}(C,C), \; V_{SSG}^{i}(D,D) < V_{SSG}^{i}(D,C). \tag{5.34}$$

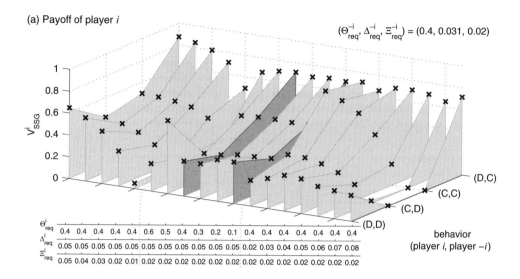

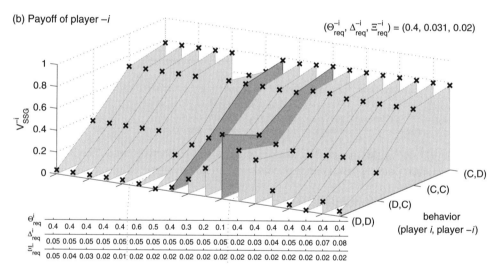

Figure 5.22 The payoffs in SSGs of two players depending on their behavior. The QoS requirement of player i is varied while the opponent's requirement is fixed: (Θ_{req}^{-i}, Δ_{req}^{-i}, Ξ_{req}^{-i}) = (0.4, 0.031, 0.02). The dark gray scenarios are considered in the rest of the section and are transferred to Tables 5.6(b) and (c). Reproduced with permission from L. Berlemann, G. R. Hiertz, B. Walke, and S. Mangold, Cooperation in Radio Resource Sharing Games of Adaptive Strategies. In: IEEE 60th Vehicular Technology Conference, VTC2004-Fall, Los Angeles CA, USA, 26–29 September 2004, 3004–3009. © 2004 IEEE

These equations are based on the assumption that an opponent's defection (D) implies a reduced payoff if player i is cooperating (C) or defecting (D) respectively. Note from player $-i$'s point of view, this assumption leads to:

$$V_{SSG}^{-i}(D,C) < V_{SSG}^{-i}(C,C), \quad V_{SSG}^{-i}(D,D) < V_{SSG}^{-i}(C,D). \tag{5.35}$$

The applicability of Equation (5.34) can be seen, for example, in Figure 5.22(a), by comparing the second row (C, D) of the payoff values with the third row (C, C) and the first row (D, D) with the fourth row (D, C). The classification fails only for $\Delta^i_{req} = 0.07$ and $\Delta^i_{req} = 0.08$, where $V^i_{SSG}(C, D)$ and $V^i_{SSG}(C, C)$ are nearly equal and zero.

5.5.8 Equilibrium Analysis

The outcome of an SSG, namely the payoff as defined in Equation (5.13), depends on the observed QoS parameters, which can be determined by the players through the analytic model of Mangold (2003). It is interesting to analyze whether or not these outcomes are steady and/or payoff maximizing. Therefore, in an SSG of rationally acting players, the existence of a best-response action on the expected opponent's action has to be considered. In addition, the uniqueness and stability of such an action is of interest for the players' decision making process regarding which action to take. A commonly used solution concept for the question of which action should be selected in an SSG is the Nash[1] Equilibrium solution concept (Nash, 1950).

5.5.8.1 Nash Equilibrium

In general, a Nash Equilibrium (NE) is a profile of strategies such that each player's strategy is a best response to the other player's strategy. Here, in the context of the SSG the player's strategy consists of a single specific action. This action leads to an observed payoff, as outcome of the SSG. In the case of a NE, none of the players can gain a higher payoff by deviating from their current action.

NEs are consistent predictions of how the game will be played. In the sense that if all players predict that a particular NE will emerge then no player has the incentive to play differently. Thus, an NE, and only an NE, can have the property that the players can predict it; predict that their opponents predict it, and so on. The NE is a value for the game's stability and there is no incentive (in terms of potentially higher payoffs) to leave, as illustrated below. If it is successfully established, the game has a steady and therefore predictable outcome. From the technical point of view, the observed QoS parameter in a NE can be seen as a lower limit for the QoS that can be guaranteed in a competition scenario of rational players.

5.5.8.2 Nash Equilibrium in Single Stage Games

The bargaining domain, as comparison of the players' payoffs from an SSG, is shown in Figure 5.23 in order to illustrate game scenario I (Table 5.6b). Starting with a game-wide behavior of cooperation (C, C), both players have the incentive to deviate from cooperation to gain a higher payoff. In the case of player $-i$ first leaving the cooperation (marked with '①' in the figure, gray dotted line) and deviating to (C, D), player $-i$'s payoff is increased, while player i observes a reduced payoff compared to the origin of (C, C). Consequently, player i decides to save (rescue) his payoff, and to punish the opponent for changing his behavior to defection (②). The resulting payoff reduction in (D, D) for player $-i$ stimulates his return to cooperation (③). As player i has no incentive, i.e. an expected higher payoff, to leave (D, C), a

[1] John F. Nash (*1928), mathematician, Nobel Price in economic science 1994.

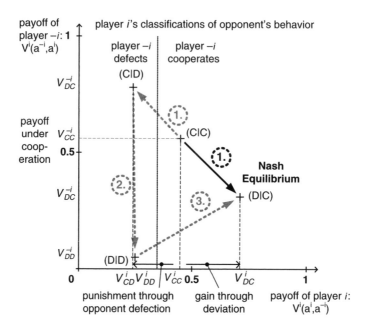

Figure 5.23 Bargaining domain of game scenario I of Table 5.6(b). Originated in a game-wide cooperation, a unique Nash equilibrium is reached. Here, the Nash equilibrium is in (D, C). Reproduced with permission from L. Berlemann, B. Walke, and S. Mangold, 'Behavior Based Strategies in Radio Resource Sharing Games,' in Proc. of 15th IEEE International Symposium on Personal, Indoor and Mobile Radio Communications, PIMRC 2004, Barcelona, Spain, 5–8 September 2004. © 2004 IEEE

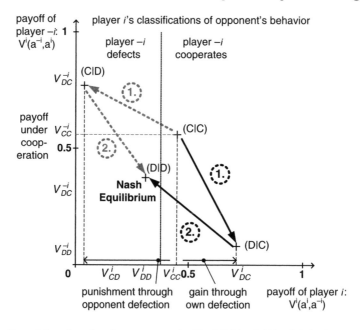

Figure 5.24 Bargaining domain of game scenario II of Table 5.6(c). Originating in a game-wide cooperation, a unique Nash equilibrium is reached. Here, the Nash equilibrium is in (D, D)

stable point is reached. The same applies for the case of player i deviating first from cooperation (①, solid line). Thus, in this specific game scenario, (D, C) is a stable point where neither
player can gain a higher payoff and has therefore no intention of deviating: (D, C) is an NE.

The alternative game scenario II (Table 5.6c) is illustrated in the bargaining domain
of Figure 5.24. the 'Prisoner's dilemma' has its unique NE in the behaviors of '(D, D)'.
Originating in game-wide cooperation (C, C), both players switch to defection in the case
of opponent defection and remain there.

5.5.8.3 Pareto Efficiency

A central microeconomic concept for judging outcomes of a game is the *Pareto efficiency*:
an SSG outcome is called Pareto efficient if neither player can gain a higher payoff without
decreasing the payoff of at least one other player. A non-Pareto efficient situation is not the
preferable outcome for a stage because a rational player could improve his payoff without
changing the game model or the situation of the other participating players. This consideration
of the opponents' payoff extends the concept of an NE where only the individual payoff of a
player is measured.

The Pareto efficiency enables a judgment of game outcomes, especially in the case of multiple NEs. NEs are stable and predictable points of operation in SSG, to which the players
adjust in case they are acting rationally. It has been shown by Mangold (Mangold, 2003)
that at least one NE exists in the SSG. If the NE is unique, the corresponding actions can
be predicted by the players as a result from rational behavior. The NE is referred to as
Pareto efficient if none of the players can improve his payoff without decreasing the payoff of any of the other players. Thus, the radio resource is optimally used by the spectrum
sharing players.

In general, there may exist action profiles in SSGs that lead to higher payoffs as compared
with the SSG outcome in the NE. In such a case, the NE is not Pareto efficient and actions
by all players may optimize the payoffs of all players. In the case of multiple NEs, the Pareto
efficiency is a criterion for assessing these NEs in order to find the most preferable one. The
bargaining domain in the next section further illustrates the judgment of game outcomes with
the help of the Pareto efficiency.

5.5.8.4 Bargaining Domain

The *bargaining domain* in Figure 5.25 contains a subset of all possible SSG outcomes,
by means of players' payoffs ($V^i|V^{-i}$), corresponding to an action pair of players i and
$-i$ ($a^i|a^{-i}$). Here, the actions belong to a discrete set. The corresponding SSG outcomes,
i.e., payoff pairs, are calculated with the help of the analytic model from Section 5.5.6. The
bargaining domain supports the judgment of potential SSG outcomes. Depending on the play-
ers' QoS requirements ($\Theta^i_{req}, \Delta^i_{req}, \Xi^i_{req}$) and ($\Theta^{-i}_{req}, \Delta^{-i}_{req}, \Xi^{-i}_{req}$), none, a unique NE or several
NEs can be found. Here, the game scenario has a unique NE with player i demanding a
throughput of 0.5 and a period length of 0.032, while the opponent $-i$ demands 0.54 and 0.03
respectively. The corresponding utilities are 0.913 and 0.872, as depicted in Figure 5.25. This
NE, which is not Pareto efficient, can be considered as a minimum for the reachable utilities
of both players. In this way a lower but nevertheless predictable limit for the support of QoS

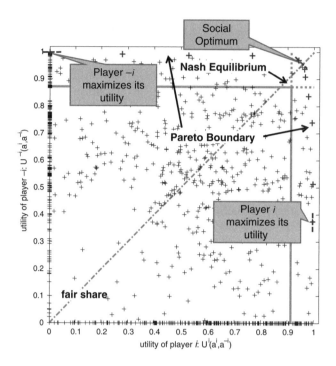

Figure 5.25 The bargaining domain of a game scenario. Each cross marks the players' utilities of an SSG and the players' actions are varied. The Pareto-efficient outcomes define the Pareto boundary. Three special Pareto efficient outcomes are highlighted. Reproduced with permission from S. Mangold, L. Berlemann, and B. Walke, Equilibrium Analysis of Coexisting IEEE 802.11e Wireless LANs. In: 14th IEEE Conference on Personal, Indoor and Mobile Radio Communications, PIMRC 2003, Beijing, China, 7–10 September 2003, 321–325. © 2003 IEEE

is given. A further analysis indicates that this unique NE is reached as a steady outcome of a Multi Stage Game (MSG) if both players follow the best response behavior.

The single NE enables a definition of a Pareto frontier (Mas-Colell *et al.*, 1997). The Pareto frontier marks the reachable outcomes under a best response behavior. All outcomes outside the Pareto frontier are characterized by Pareto domination of the NE. The game scenario illustrated in Figure 5.25 has a multitude of Pareto efficient outcomes, which border the payoffs from all SSGs at the right and upper corner of the bargaining domain. The payoff pairs with a maximum payoff for each player that lie on this border line, are depicted in Figure 5.25. For player $-i$ it is located in the upper left area of the bargaining domain and for player i in the lower right area. A special Pareto efficient outcome is located in the upper right area with the longest distance to the origin of the bargaining domain. There, both players gain, contrary to the other two Pareto efficient outcomes, a higher payoff than in the NE. As a result, both players can improve their payoffs through interaction, compared with utilities in the NE. This interaction is referred to as cooperation to the benefit of all players and is the motivation for the focus on MSGs in the next section. Additionally, the sum of both players' payoffs has, in the upper right corner, the highest value of all in the bargaining domain. It can therefore be regarded as a social optimum.

5.5.9 Multi Stage Game Model

Repeated SSGs from an MSG. Concrete, the MSG can be characterized as follows:

- the MSG consists of a finite number of stages, and the end of the MSG is unknown to players (which allows the game to be modeled as infinitely repeated games);
- players may observe different outcomes and maintain their own local information base about the status of the game. Information is non-symmetrically distributed between players;
- the actions are taken (behaviors are selected) at the beginning of each stage;
- there are no mixed strategies, hence, there is no probability distribution associated with the set of available behaviors. A player takes one single action at the beginning of each SSG;
- at each stage, players obtain a history H^n of observed outcomes of the past stages.

Technical restrictions, such as the battery power of a mobile terminal for example, limit the duration of the MSG, hence there are no games of infinite duration. A finitely repeated game with known end is solved with backward induction, that is, from the known end, the outcome of the last stage can be calculated. Based here on the outcome of the previous stage, all other outcomes back to the beginning of the game are determinable. This backward induction is not possible in the case of the end of an MSG being unknown to the players. Here, it is assumed that players do not know when the interaction ends; hence the MSG is a finitely repeated game with an unknown but existing end. Nevertheless, it can be assumed that the MSG has no limited time horizon and the MSG may be regarded as infinite:

'A model with infinite horizon is appropriate if after each period the players believe that the game will continue for an additional period.' (Osborne and Rubinstein, 1994)

For these reasons, it can be legitimately assumed that the MSG is infinite.

5.5.10 Discounting of Future Payoffs

When selecting how to access the medium, players take into account the expected results (the expected utilities) of the instantaneous stage, but should also take into account the effects of their decisions on the payoffs of future stages. This is usually expressed in game theory through weighting the stages. Players give present payoffs a higher weight than potential payoffs in the future, because of the uncertainty of those future results. A known approach to model this weighting of the future is to discount the payoffs for each future stage of a game. Therefore, a discounting factor λ, $0 < \lambda < 1$, can be defined that reflects in the present stage the value of future payoffs of following stages.

A λ value near unity implies that future utilities are considered similarly to the payoff of the current stage. Thus, the player tends to cooperate to enable a high long-term payoff. Alternatively, a player with a λ near zero only has his focus on the present payoff and completely neglects potential future payoffs, thus resulting in non-cooperating defection.

The exact value of this discounting factor λ is determined by the applications that are supported by the player (as is also the case for the shape of the utility function). It is derived from the technical requirements of the QoS traffic types resulting from the applications supported

by the players. See Table 5.7(a) for different QoS traffic types and the corresponding discounting factors λ determined through QoS sensitivities from Dutta-Roy (2000). A definition
near to the 802.11e standard is introduced in Table 5.7(b), but is not used in this section. The
QoS traffic types and the corresponding discounting factors λ are determined through different access categories of IEEE 802.11e (IEEE, 2004a), which are derived from the technical
report IEEE 802.1D (IEEE, 1998).

Table 5.7   Discount factors of different traffic types

(a) based on Dutta-Roy (2000)

Traffic type	Sensitivities				λ
	Throughput (Θ)	Loss	Delay (Δ)	Jitter (Ξ)	
e-Mail	Low	High	Low	Low	0.5
Browsing	Low	Medium	Medium	Low	0.6
Voice	Very low	Medium	High	High	0.75
Video Conferencing	High	Medium	High	High	0.9
Multicasting (Network Control)	High	High	High	High	1

(b) based on IEEE 802.11e which is derived from the technical report IEEE 802.1D

Traffic type	802.1D Priority	802.11e Access category	λ
Best effort	0	0	0.25
Excellent effort	3	1	0.5
'Video,' <100 ms latency and jitter	5	2	0.75
'Voice,' <10 ms latency and jitter	6	3	0.9

Under the assumption that a large number of stages is left to be played, player i's payoff
V^i_{MSG} of an MSG is defined as the sum over its payoffs V^i_n of stage n discounted with λ^i
(Berlemann, 2003; Mangold, 2003):

$$V^i_{MSG} = \sum_{n=0}^{\infty} \left(\lambda^i\right)^n V^i_n = \frac{1}{1-\lambda^i} V^i_n, \; if \; V^i_n = const. \; \forall n = 0 \ldots \infty \tag{5.36}$$

5.5.11 Strategies

In MSGs, strategies determine the behaviors for each individual SSG. Players try to optimize
their payoff by applying adequate strategies. Using a state model as defined in Figure 5.26(a),
a strategy describes a player's alternatives. Each state represents a certain behavior. A strategy
also models those circumstances under which a transition from one state to another hap-
pens; hence, it models the decision making and learning. This section only allows what is
referred to in game theory as a 'pure' strategy (Fudenberg and Tirole, 1998), where players have to choose one specific behavior for each stage, and cannot perform soft decisions

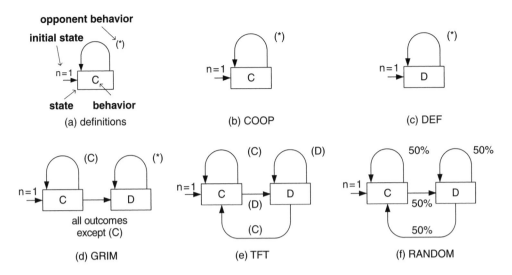

Figure 5.26   State machines for modeling strategies as defined in (a). Static strategies: strategy of cooperation (b) and defection (c). Trigger strategies: GRIM strategy (d), tit-for-tat strategy (e) and RANDOM strategy (f). Reproduced with permission from L. Berlemann, G. R. Hiertz, B. Walke, and S. Mangold, 'Strategies for Distributed QoS Support in Radio Spectrum Sharing,' in Proc. of IEEE International Conference on Communications, ICC 2005, Seoul, Korea, 16–20 May 2005. © 2005 IEEE

by assigning probabilities to different state transitions. Following Osborne and Rubinstein (1994), strategies can be interpreted as social norms in repeated interaction:

> 'Social norms are isolated types of strategies that support in any game mutually desirable and thus stable utilities'

In other words, strategies enable QoS support independently from the opponent's strategy and QoS requirements. This section mainly discusses MSGs of two players. Thus the focus is on the strategy pairs of these two players, denoted as (strategy of player i | strategy of player $-i$). Further, this section distinguishes between static and dynamic (trigger) strategies, as explained below.

5.5.11.1 Static Strategies: Cooperation and Defection

Static strategies are the continuous application of one behavior without regarding the opponent's strategy. In static strategies, there is no state transition, and the state model contains one single state. In the present approach, the set of available static strategies is reduced to two: The cooperation strategy (COOP) is characterized by cooperating every time, independently of the opponent's influence on a player's payoff. The COOP strategy is to the benefit of a player if the opponent cooperates as well. Figure 5.26(b) illustrates the simple strategy of following a cooperative (C) behavior. Equivalently to the COOP strategy, the defection strategy (DEF) consists of a permanently chosen behavior of defection (D). Figure 5.26(c) illustrates the DEF strategy as a state machine.

5.5.11.2 Dynamic (trigger) Strategies: GRIM and Tit-For-Tat

Trigger strategies are well known in game theory (Fudenberg and Tirole, 1998; Osborne and Rubinstein, 1994). A trigger strategy is a dynamic strategy where the transition from one state to another state is event driven; an observed event triggers a behavior change in a player. Depending on the number of states (the number of behaviors a player may select), a large number of trigger strategies is possible. For the sake of simplicity, the familiar Grim (GRIM) and Tit-For-Tat (TFT) trigger strategies are applied in the following. A player with a GRIM strategy punishes the opponent for a single deviation from cooperation with a defection forever. A player applying this strategy may be referred to as an unforgiving player. The initial state of the GRIM strategy, selected at the first stage of the MSG, is, however, cooperation. The player cooperates as long as the opponent cooperates, and the transition to defection is triggered by the opponent's defection. See Figure 5.26(d) for an illustration of the state machine of the GRIM strategy. The TFT strategy selects cooperation as long as the opponent is cooperating, similarly to the GRIM strategy, and also with cooperation in the initial stage. An opponent's defection in stage N triggers a state transition. This transition is punished by defection in the following stage $(N + 1)$, as illustrated in Figure 5.26(e). However, in contrast to the GRIM strategy, TFT changes back as soon as the opponent cooperates again. The TFT strategy is well known in game theory and in social science. The advantage of the TFT strategy is that on the one hand it motivates opponent players to cooperate (because of the potential punishment), and on the other hand, its robustness when applied in non-cooperative environments, where opponent players often defect.

5.5.11.3 RANDOM Strategy

We also want to analyze how the different strategies perform when applied to purely random behavior. To analyze whether a random play or a deterministic, predictable play that usually results in a stable course of the game is to the advantage of a player, we introduce the dynamic strategy RANDOM. This strategy, as shown in Figure 5.26(f), results in uniformly distributed behaviors, 50 % cooperation (C) and 50 % defection (D), regardless of what behavior the opponent player may select. This RANDOM strategy is compared in Berlemann *et al.* (2004b) to a DEF5of10 strategy, not depicted here, that results in five stages of C followed by five stages of D with, again, five stages of C thereafter, and so on. The DEF5of10 is a variation of the RANDOM strategy given above and indicates that, arbitrarily, many additional strategies exist.

5.5.11.4 Adaptive Strategies

The opponent's strategy is unknown to a player, but nevertheless, the resulting behavior can be observed and classified. Consequently it is preferable to have flexible strategies, which enable the player individually to adapt his own strategy during an MSG to those of the opponent. The players have, in general, the intention that all players within the game can find their optimal strategy to achieving a better outcome for all. The already introduced strategies form a framework for further refinement of the dynamic interaction. These necessarily more common strategies are called adaptive strategies, which are derived from the trigger strategies above, and paying tribute to the limited information about the opponent within the MSG. Depending on the player's payoff table, which is unknown to the opponent, a player may adapt his strategy to achieve the strategy's intention in an optimal way.

An example for an adaptive strategy is the *n-Tit for n-Tat* (nTFnT) strategy by Anatol Rapoport (Axelrod, 1984), as is shown in the state machine of Figure 5.27. The *n*-tit part reflects the strength (i) while the *n*-tat part stands for the tolerance (ii). The relative difference between the payoffs in the SSG, corresponding to both players' behavior, is decisive for the choice of the strategy. In the face of the manifoldness of the game scenarios as depicted in Figure 5.22 is the nTFnT strategy, suitable for an individual adaptation to the opponent's strategy in order to enforce a stable course of the MSG for a better capability of QoS support.

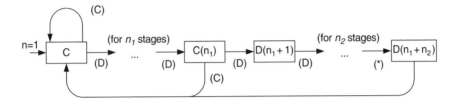

Figure 5.27 State machine of the adaptive strategy *n*-tit for *n*-tat. Reproduced with permission from L. Berlemann, G. R. Hiertz, B. Walke, and S. Mangold, Cooperation in Radio Resource Sharing Games of Adaptive Strategies. In: IEEE 60th Vehicular Technology Conference, VTC2004-Fall, Los Angeles CA, USA, 26–29 September 2004, 3004–3009. © 2004 IEEE

More sophisticated approaches to learning in games are discussed for example in Fudenberg and Levine, (1998).

5.5.12 Nash Equilibrium in Multi Stage Games

This section analyzes MSGs under the microeconomic aspects of welfare, constant requirements, evolving demands, and the resulting payoffs. Throughout the course of MSGs, players attempt to optimize their payoffs by changing their behaviors when needed. Each player follows a strategy to determine which behavior to select in a stage. In evaluating the payoff, MSG NEs for optimized QoS support are determined. Initially, the following section answers in general the question of whether or not a random play that implies many changes in behavior, or an aimed interaction resulting into a stable course of the game, is to the advantage of a player.

To guarantee QoS continuously by achieving predictable outcomes, even at a minimum predictable level, the players intend to establish a steady state. In this state, behavior does not change significantly, and future outcomes are predictable. Typically, the players attempt to influence this steady state to their advantage. The NE of an SSG implies best response actions with satisfying outcomes and is the basis for analyzing steady states in MSGs.

In extending the scope to multiple stages, the level of potential interaction increases. The players have to consider future stages in their decision as to which action to choose for a given stage. Therefore, static and dynamic trigger strategies are considered to improve the SSG outcomes in steady state. In this context, the concept of the NE has to be extended to MSGs. Up to here, this section has considered Nash equilibria for SSGs, now this section concentrates on finding Nash equilibria for MSGs. The NE of an MSG is based on the players' strategies for the MSGs, as opposed to actions for the SSG. A pair of strategies is not a NE if any strategy can be found under whose application either player would gain a higher payoff. No player can gain a higher payoff in deviating from a strategy for a single stage and returning thereafter to this strategy. A strategy pair is an NE if no strategy can be found that is more preferable for any player. An NE of strategies in this section corresponds to the classical understanding of an NE in game theory (Fudenberg and Tirole, 1998; Osborne and Rubinstein, 1994).

Trigger strategies may lead to higher payoffs for all players because they allow better mutual adaptation of behaviors. Therefore, the existence of NEs in such MSGs and the corresponding trigger strategies are analyzed below.

5.5.12.1 Evaluation of Tit-for-Tat Strategy

In this section the focus is on MSGs with two players in which (i) both players prefer the TFT strategy, and (ii) both players may seek incentives for alternative strategies. Player i prefers to switch from TFT to DEF if $V^i(DEF|TFT) > V^i(TFT|TFT)$, while player $-i$ continues to apply the TFT strategy. In choosing the DEF strategy, player i gains a higher payoff V_{DC}^i at one stage, and in the subsequent stages, after player $-i$ switches from cooperation to defection too, V_{DD}^i for the rest of the game. Thus the payoff inequality is solved, discounting Equation (5.36) to:

$$V_{DC}^i + V_{DD}^i \frac{\lambda^i}{1 - \lambda^i} > V_{CC}^i \frac{1}{1 - \lambda^i} \tag{5.37}$$

In isolating λ^i, it can be derived that (TFT|TFT) results for:

$$\lambda^i > \frac{V_{DC}^i - V_{CC}^i}{V_{DC}^i - V_{DD}^i} \tag{5.38}$$

in higher payoffs than (DEF|TFT). As illustrated in the bargaining domain of Figure 5.23, the term $V_{DC}^i - V_{CC}^i$ can be regarded as the gain of player i from leaving the cooperation, while $V_{DC}^i - V_{DD}^i$ is the punishment that consequently follows the defection in TFT strategies.

Next, a DEFk strategy is defined, which implies k defections in stage L, and cooperation thereafter. The strategies DEF and DEF1 form a border for more unlimited strategies between these two. A player has to compare the single deviation gain on the one hand to the consequential punishment for one stage by the opponent with the TFT strategy in the next stage on the other hand. Therefore, player i prefers to switch from TFT to DEF1 if

$$V_{DC}^i \left(\lambda^i\right)^L + V_{CD}^i \left(\lambda^i\right)^{L+1} > V_{CC}^i \left(\lambda^i\right)^L + V_{CC}^i \left(\lambda^i\right)^{L+1} \tag{5.39}$$

Accordingly, (TFT|TFT) results in higher payoffs compared with (DEF1|TFT) for:

$$\lambda^i > \frac{V_{DC}^i - V_{CC}^i}{V_{CC}^i - V_{CD}^i} \tag{5.40}$$

The term $V_{CC}^i - V_{CD}^i$ implies the payoff reduction for player i in case of an initial opponent deviation or an executed punishment as reaction. For λ^i of player i within the restriction of Equation (5.38), (TFT|TFT) results higher payoffs than (DEF|TFT). Game scenario I in Table 5.6(b) leads, for player 1, to a value of $\lambda^i \geq 0.690$. Equation (5.40) restricts λ^i more, leading in the same scenario to $\lambda^i \geq 2.05$. Consequently, there is no $0 < \lambda^i < 1$ for which player 1 would not prefer to switch to (DEF1|TFT). Alternatively, with game scenario II of Table 5.6(c), in accordance with Equation (5.40) player 1 would choose (TFT|TFT) for a $\lambda^i \geq 0.861$.

Based on the definition of an NE from Section 5.5.8.1, the most restrictive value for λ^i is decisive for (TFT|TFT) being an NE. As the example illustrates, the strategy DEF1 implies the highest temptation to defect and can be thus regarded in this way as lower limit for λ^i under which (TFT|TFT) is an NE.

Figure 5.28 illustrates the above results from above. The analytic MSGs are solved using the mathematical model from Section 5.5.6, while the simulation results are from the Matlab-based simulation tool YouShi, introduced in Appendix B. The dashed gray horizontal line marks the payoff from game-wide cooperation. The different strategies of player i imply a specific number of stages in which player i deviates from cooperation during an MSG. The opponent has a constant TFT strategy. The stages of defection are increased on the x-axis from 0 to 10 and the corresponding strategies are also indicated. The resulting payoffs of the MSG outcomes V^i of player i, with the payoff tables from the game scenarios I and II are depicted. The resulting payoffs are normalized to the payoff from game-wide cooperation. The potential strength of punishment through the opponent, given by the specific game scenario, is decisive for player i in determining which strategy to choose. The strategies DEF1 and DEF2 lead to higher payoffs compared with the case of game-wide cooperation. Game theory calls DEF1 and DEF2 dominating strategies. Player i has a discounting factor of 0.9. For increasing λ^i the payoffs from defection strategies are reduced due to the growing relevance of expected punishment.

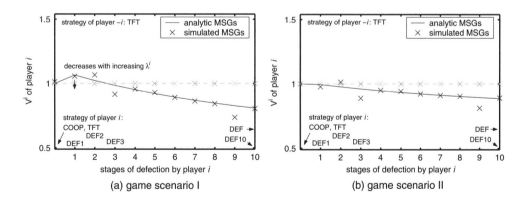

Figure 5.28 MSG payoffs of player i, for two game scenarios (a) and (b) of different QoS requirements. Reproduced with permission from L. Berlemann, B. Walke, and S. Mangold, 'Behavior Based Strategies in Radio Resource Sharing Games,' in Proc. of 15th IEEE International Symposium on Personal, Indoor and Mobile Radio Communications, PIMRC 2004, Barcelona, Spain, 5–8 September 2004. © 2004 IEEE

5.5.12.2 Evaluation of GRIM Strategy

This section focuses on the GRIM strategy. Player i, playing against an opponent who also applies the GRIM strategy prefers to switch from GRIM to DEF if V^i $(DEF|GRIM)$ > V^i $(GRIM|GRIM)$. In a game of (GRIM|GRIM), both players are cooperating during the complete game. In applying a DEF strategy, player i would gain in the first stage a deviation gain because player $-i$ begins with cooperation. After the first stage, player $-i$ defects for the rest of the game, following his GRIM strategy, and player i receives a reduced payoff because both players are defecting. The payoff equation is then given by:

$$V^i_{DC} + V^i_{DD} \frac{\lambda^i}{1 - \lambda^i} > V^i_{CC} \frac{1}{1 - \lambda^i} \tag{5.41}$$

Thus (DEF|GRIM) leads to a higher payoff than (GRIM|GRIM) for

$$\lambda^i > \frac{V^i_{DC} - V^i_{CC}}{V^i_{DC} - V^i_{DD}} \tag{5.42}$$

and is in this case an NE. This is the same restriction for λ^i as in Equation (5.38), resulting from the comparison of DEF and TFT.

5.5.12.3 Existence of Nash Equilibria

The general existence of NEs in MSGs depends on the relationship between the individual payoffs of each player and the possibility of those players to influence each other. Here the focus is on MSGs with a restriction of $V^i_{DC} - V^i_{DD} > V^i_{CC} - V^i_{CD}$. In other words, the payoff loss due to a defection of the opponent is smaller than the punishment through the opponent after an own defection. Hence, the case of Equation (5.40) is more restrictive than Equation (5.38) by means of:

$$\frac{V^i_{DC} - V^i_{CC}}{V^i_{DC} - V^i_{DD}} < \frac{V^i_{DC} - V^i_{CC}}{V^i_{CC} - V^i_{CD}} \tag{5.43}$$

In this way a lower limit for λ^i of player i is defined under which an NE from player i's point of view is established. Furthermore is it easier to sustain a GRIM strategy pair as an NE than a TFT pair because of Equation (5.43).

5.5.12.4 Evaluation of Game Scenarios

Table 5.8(a) and (b) compare the strategies with the help of the MSG outcomes. The players have QoS requirements corresponding to the game scenarios of Table 5.6(b) and (c) respectively. The outcomes of the MSGs, depending on the players' strategy, are calculated with the help of the discounted payoffs from Equation (5.36). The payoffs are normalized to the MSG outcome of game-wide cooperation, the MSGs are played over 10 stages and both players have a discounting factor of $\lambda^i = \lambda^{-i} = 0.9$. The NEs of the MSGs are highlighted in gray in both tables. Analogous to the SSG the strategy pair of (DEF|COOP) is the unique NE of the MSG of scenario I (Table 5.8a). For higher discounting factors, satisfying Equation (5.40) for both players, the strategy pair (TFT|TFT) would be the emerging NE as it is the case in scenario II (Table 5.8b). There, strategies that imply a game-wide cooperation lead to satisfying steady outcomes; the strategy pairs (GRIM|GRIM), (GRIM|TFT), (TFT|GRIM) and (TFT|TFT) are the Pareto efficient NEs of this MSG. Analogous to the SSG, the strategy pair of (DEF|DEF) is still a Pareto inefficient additional NE. In these two example game scenarios, the GRIM strategy dominates the TFT strategy and leads always to equal or higher payoffs than TFT, because in these scenarios the punishment of the opponent simultaneously implies a significant increase of the own payoff.

Table 5.8 Discounted MSG payoffs of both players depending on the players' strategy. The players have QoS requirements corresponding to game scenario I (a) and game scenario II (b). Reproduced with permission from L. Berlemann, B. Walke, and S. Mangold, 'Behavior Based Strategies in Radio Resource Sharing Games,' in Proc. of 15th IEEE International Symposium on Personal, Indoor and Mobile Radio Communications, PIMRC 2004, Barcelona, Spain, 5–8 September 2004. © 2004 IEEE

(a) Game scenario I

Pl.1↓ Pl.2→	COOP	DEF1	DEF	GRIM	TFT
COOP	1.00, 1.00	0.91, 1.09	0.60, 1.39	1.00, 1.00	1.00, 1.00
DEF1	1.17, 0.90	0.91, 0.80	0.61, 1.10	0.86, 1.20	1.10, 0.97
DEF	1.78, 0.55	1.52, 0.45	0.63, 0.09	0.88, 0.19	0.88, 0.19
GRIM	1.00, 1.00	1.51, 0.74	0.62, 0.38	1.00, 1.00	1.00, 1.00
TFT	1.00, 1.00	1.05, 1.01	0.62, 0.38	1.00, 1.00	1.00, 1.00

(b) Game scenario II

Pl.1↓ Pl.2→	COOP	DEF1	DEF	GRIM	TFT
COOP	1.00, 1.00	0.86, 1.06	0.10, 1.39	1.00, 1.00	1.00, 1.00
DEF1	1.12, 0.87	0.96, 0.95	0.20, 1.28	0.36, 1.20	0.99, 0.92
DEF	1.78, 0.14	1.62, 0.23	0.78, 0.68	0.93, 0.60	0.93, 0.60
GRIM	1.00, 1.00	1.52, 0.33	0.67, 0.79	1.00, 1.00	1.00, 1.00
TFT	1.00, 1.00	0.97, 0.94	0.67, 0.79	1.00, 1.00	1.00, 1.00

5.5.13 QoS Evaluation of Strategies

The previous section analyzed strategies in MSGs on the basis of payoffs (as a single summarizing value for multiple QoS parameters). This section increases the detail of evaluation by considering the support of these QoS parameters during an MSG. The game model from above and the basic IEEE 802.11e access mechanisms to a shared resource are evaluated with the help of the Matlab-based simulator YouShi introduced in Appendix B. A player instance as part of the Station Management Entity (SME) realizes the game approach of this section in the protocol stack of IEEE 802.11e (IEEE, 2004a).

5.5.13.1 Strategies for QoS Support in Multi Stage Games

In the following section the QoS capabilities of the strategies introduced in Section 5.5.11 are evaluated. Multiple MSGs with varying strategies for both players are evaluated in the two game scenarios from above and the QoS requirements of player i are different from one

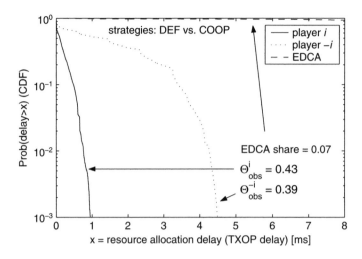

Figure 5.29 MSG example. Observed throughput (Θ) and TXOP delay of both players in the presence of less prior EDCA traffic, when player i has a DEF strategy while player $-i$ follows the COOP strategy. Defection leads to better QoS outcomes than does cooperation. Reproduced with permission from L. Berlemann, G. R. Hiertz, B. Walke, and S. Mangold, 'Strategies for Distributed QoS Support in Radio Spectrum Sharing,' in Proc. of IEEE International Conference on Communications, ICC 2005, Seoul, Korea, 16–20 May 2005. © 2005 IEEE

scenario to the other while player $-i$ has fixed ones. Game scenario I from Table 5.6(b) is evaluated from the perspective of player i in Figure 5.30 and for player $-i$ in Figure 5.31. Each strategy pair has a corresponding course in the MSG and is noted as (strategy of player i | strategy of player $-i$), for example (GRIM|TFT). Such a strategy pair results in specific combinations of behaviors in the SSGs of the MSG, and is noted as (behavior of player i, behavior of player $-i$); for example (C, C). The QoS outcomes in MSGs of various strategies are summarized. The MSG outcomes from one MSG example are depicted in Figure 5.29. The parts of Figure 5.30 illustrate the observed QoS of player i: The achievable throughput Θ^i_{obs}, which is given as fraction of total capacity, and the complementary Cumulative Distribution Function (CDF) of the Transmission Opportunity (TXOP) allocation delays, denoted as Ξ^i_{obs}. Player i's strategy is fixed in each part, while opponent $-i$ alters his strategy. The observed throughputs Θ^i_{obs} and TXOP allocation delays Ξ^i_{obs} of player i are evaluated over the outcomes of 400 stages of two player MSGs. The gray lines mark the 98 % percentile of the TXOP delay, and 98 % of the allocations observe a delay of less than the corresponding delay value at the curve's crossing point with this line.

The players have normalized QoS requirements corresponding to game scenario I of $(\Theta^i_{req}, \Delta^i_{req}, \Xi^i_{req}) = (0.4, 0.05, 0.02)$ and $(\Theta^{-i}_{req}, \Delta^{-i}_{req}, \Xi^{-i}_{req}) = (0.4, 0.031, 0.02)$. Parameter Δ^i_{req} determines the period between two consecutive TXOP allocation attempts. A third participating player represents the background traffic and the contention-based medium access, the Enhanced Distributed Channel Access (EDCA) as introduced in Appendix B of both Wireless Local Area Networks (WLANs). This player is not considered in this evaluation; however, Figure 5.29 shows the EDCA TXOP delays.

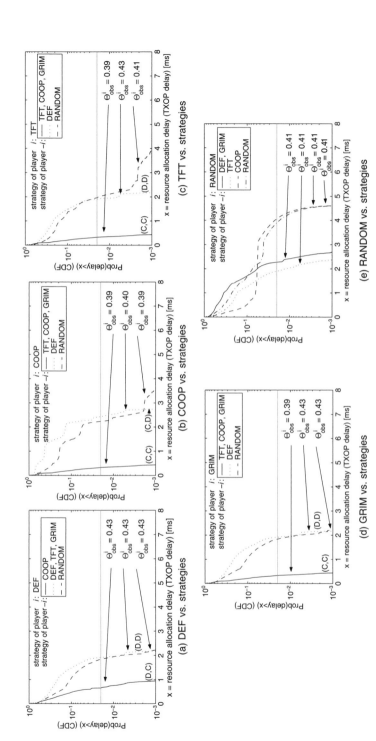

Figure 5.30 CDFs of resource allocation delays and observed mean throughputs resulting from MSGs of player i in game scenario I. Player i has a fixed strategy while the opponent's $-i$ strategy is varied. Reproduced with permission from L. Berlemann, G. R. Hiertz, B. Walke, and S. Mangold, 'Strategies for Distributed QoS Support in Radio Spectrum Sharing,' in Proc. of IEEE International Conference on Communications, ICC 2005, Seoul, Korea, 16–20 May 2005. © 2005 IEEE

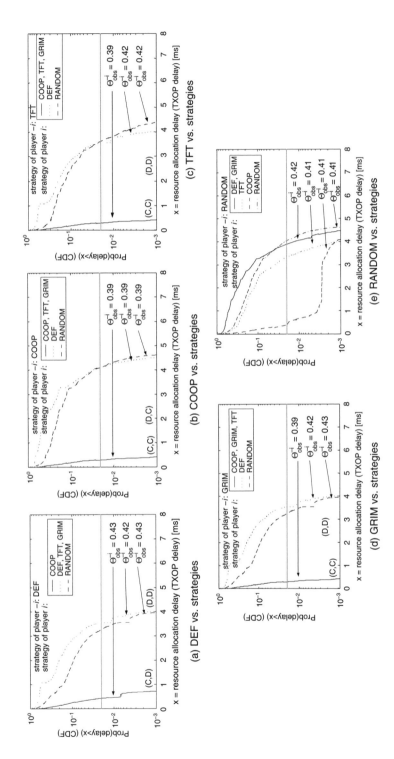

Figure 5.31 CDFs of resource allocation delays and observed mean throughputs resulting from MSGs of player $-i$ in game scenario II. Player $-i$ has a fixed strategy while opponent's i strategy is varied

In the MSG example in Figure 5.29, player i follows the DEF strategy while opponent $-i$ applies the strategy COOP. Player i exceeds his required throughput $\Theta^i_{obs} > \Theta^i_{req} = 0.4$ contrary to the opponent, who misses his requirement $\Theta^{-i}_{obs} < \Theta^{-i}_{req} = 0.4$. In terms of allocation delays, player i with his DEF strategy observes much shorter TXOP delays than the cooperating opponent. These strategies imply permanent behavior of cooperation for player i and defection for player $-i$, i.e. (C, D).

The observed QoS of player i resulting from the DEF strategy is illustrated in Figure 5.30(a). The observed throughputs are for all opponent strategies higher than the required one and thus this QoS requirement is fulfilled. In general, the throughput is not the critical aspect in the mutual QoS coordination within the MSGs considered here, as both players together demand 80 % of the capacity. The remaining time is allocated through the third subordinated accessing player, which has a higher demanded throughput than observed. As the DEF strategy with its permanent behavior of defection implies the SSG payoff-optimizing best response, the two players share up the medium in blocking out the third player. This is always the case if both players defect during the course of the MSG, leading to repeated SSGs of (D,D). The DEF strategy leads against COOP for 98 percent of the allocation attempts to delays less than 0.7 ms. Thus, the strategy of DEF is the best strategy compared to the others with delays of 2.0 ms.

The success of the COOP strategy is evaluated in Figure 5.30(b), where MSGs of game-wide cooperation imply the shortest delays for all players. The opponent's strategies of TFT, COOP and GRIM lead to a stable game course of repeated SSGs with (C, C), leading 98 % of the delays to values less than 0.3 ms. Both players reduce their period length, causing frequent attempts of short allocations and giving the opponent an increased opportunity for accessing the medium. The allocation attempts of a third, less prioritized, EDCA player benefit as well. As consequence, player i slightly misses his required throughput Θ^i_{obs} as he has to wait until the medium is idle, even if an allocation of the third player is ongoing. The strong DEF strategy of the opponent enforces its allocation scheme on the allocations of player i's COOP strategy, leading to the step-like shape of the curves in the CDF of (COOP|DEF). The RANDOM strategy of the opponent results into a CDF between the DEF and COOP as it consists of periods with behaviors of (C, C) as well as with (C, D).

Figure 5.30(c) illustrates the observed QoS of a TFT strategy against various opponent strategies. Following the TFT strategy, player i cooperates if the opponent $-i$ does the same, leading to a behavior of (C, C) for COOP, TFT and GRIM as opponent strategy. The tail/step of the delay CDF against a RANDOM strategy, which is observable in all parts of the figure, is reasoned as follows. The course of the MSG is unstable because of the random variation between D and C of the RANDOM strategy from player $-i$. This leads to a frequent adaptation of the best response corresponding to the defection of player i from his TFT strategy when following the behavior of the RANDOM strategy. This is done in parallel with an offset of one stage as a player needs one stage of observation before reacting to a change in an opponent's behavior.

The observed delays for the GRIM strategy are depicted in Figure 5.30(d). In this specific scenario, GRIM achieves the most reliable QoS results in comparison with the delay CDFs and observed throughputs of the other strategies. Player i is cooperating if the opponent cooperates, analogous to the TFT strategy. Contrary to the case of an opponent defection, the GRIM player defects forever and thus stabilizes at least his own allocations by not following the opponent's variations in behavior. The transition behaviors of (C, D) and (D, C) do not emerge compared with an MSG of (TFT|RANDOM). Here, in this specific scenario, avoidance of behavior transitions is obviously to the advantage of the player in terms of observed delays.

The observed QoS in applying a RANDOM strategy is illustrated in Figure 5.30(e). In general, this strategy with its frequent fluctuation in behavior, leads to adaptation processes of player i's best response and/or the opponent's best response resulting from the behavior of defection. This has unsatisfying high delays for player i as consequence. The results motivate a stable course of the MSG to enable a predictable MSG outcome as also elaborated by Berlemann *et al.* (2004b).

The parts of Figure 5.31 illustrate the observed QoS of the player $-i$ in the same game scenario I. As the focus in these parts is on player $-i$, player i is referred to as the opponent in the following discussion of the results. The achievable throughput Θ_{obs}^{-i}, and the CDF of TXOP allocation delays are, similarly to Figure 5.30, depicted. Player $-i$'s strategy is fixed in each part of the figure, while opponent i alters his strategy. Again the outcomes are evaluated over 400 stages in MSGs of two players.

A general comparison of the results from Figures 5.30 and 5.31 indicates that player $-i$ observes considerably higher delay than does player i when both players apply defective behavior. This is especially the case when the opponent follows the RANDOM or DEF strategy. This is explained by the different required period lengths of the players resulting in dissimilar allocation patters. These allocation patterns have an advantage for player i related to TXOP delay as consequence. Player i's $\Delta_{req}^{i} = 0.05$ while player $-i$ has $\Delta_{req}^{-i} = 0.031$ leading to a stronger position in the competition for spectrum access (resulting from the best response of defection) as discussed below. Thus player $-i$ heavily depends on game-wide cooperation in order successfully to support QoS. Also the RANDOM strategy, as evaluated in Figure 5.31(e), leads to insufficient delays of up to 4 ms, and also a high percentage of the TXOPs is not delayed much.

The characteristics of the RANDOM strategy applied by player i are again observable: Following the RANDOM strategy, player i switches between C and D with a 50 % change. Thus, player $-i$ can benefit in 50 % of all SSGs from i's cooperation if permanently defecting. This explains the shorter delays in upper part of the CDFs in the case of opponent i using the RANDOM strategy. Nevertheless, the frequent fluctuation in the behavior of player i leads to miscalculations of player $-i$'s best response to the defective behavior. The crossing of the delay CDFs of RANDOM and DEF is a result of this. The maximum evaluated delay is longer in the case of an opponent with the dynamic RANDOM strategy than with a static DEF strategy. This disadvantage of dynamics in MSGs is especially observable in Figure 5.31(c), where player $-i$ applies the TFT strategy.

5.5.13.2 Inter-strategy Comparison

For the quantitative comparison of the success of different strategies of a player taking account of the opponent's strategies as a whole, a summarizing value for each strategy is defined in the following. From this, the focus is on the QoS values resulting from one strategy against the generality of all opponent strategies. The focus is not on the success against a specific single strategy of the opponent. Therefore, a weighted Strategy Comparison Index (SCI) of player i is defined as:

$$SCI^i := \sum_{k=1}^{K} \underbrace{w_k \cdot 98\,\%\,percentile^i_k}_{delay} \cdot \underbrace{\left(1 + \Theta^i_{req,k} - \Theta^i_{obs,k}\right)}_{throughput} \qquad (5.44)$$

where $K = 3$ is the number of considered opponent strategies $k \in$(COOP,DEF,RANDOM) and w_k the weight of strategy k. To compare the different delay CDFs of Figures 5.30 and 5.31, this section focuses on the 98 % percentile of the resource allocation delay (TXOP delay) of player i and equal weights with $w_k = 1/K$ are assumed. The 98 % percentile is marked in the figures with a gray line. The term $1 + \Theta^i_{req,k} - \Theta^i_{obs,k}$ reflects the fulfillment of the required throughput of player i in applying the considered strategy against the opponent's strategy k. Small SCI values are better than large ones. The strategies of COOP, DEF and RANDOM are chosen as representative for all to the opponent available strategies. The SCIs for both players resulting from MSGs, analogous to Figures 5.30 and 5.31 are summarized in Figure 5.32. The parts of Figure 5.30 hence lead to the graph marked with crosses, the parts of Figure 5.31 to the graph marked with stars and the other values are the results for scenario II, respectively. The smallest SCI values indicate the strategy that is most effective against all opponent strategies and the best strategy if the opponent's strategy is unknown to the player.

For a better understanding of the results shown in Figure 5.32, the concept of strength of a player is introduced again. The strength of a player against his opponent is decisive for his capability to guarantee QoS. The strength is scenario dependent and is determined by the relationship between the players' QoS requirements, namely whether these QoS requirements imply strict constraints or robust allocations. In comparing (DEF|COOP) of Figure 5.29 and (COOP|DEF) of Figure 5.30(b), the relative allocation strength of player i is observable. Its observed allocation delays in the case of own cooperation and opponent defection are less than 2.7 ms (Figure 5.30b) while, on the other hand, opponent $-i$ observes delays up to 4.4 ms (Figure 5.30c) when cooperating while player i defects. The strength of a player is also discussed by Berlemann (2006) in the context of punishment.

In the following the results shown in Figure 5.32, corresponding to game scenario I, are discussed. The GRIM strategy is the most adequate one for both players to support QoS successfully. The best response behavior corresponding to the defection is successful against a defecting opponent. Nevertheless, this benefits player i from MSGs of game wide cooperation leading to shorter delays and so, in summary, the GRIM strategy is the most suitable one. The same goes for the results of player $-i$, although the DEF strategy has nearly the same SCI value as GRIM. The difference in the course of the MSG between the GRIM and DEF strategy is the behavior in the case of a cooperating opponent. (C, C) leads to shorter delays for player $-i$ than (C, D), as observed when comparing DEF and GRIM in Figure 5.31(a) and (d). The slightly missed throughput requirement in Figure 5.31(d) outweighs the longer delays resulting from opponent defection in the case of own cooperation in the extreme left graph of Figure 5.31(a). Thus, player $-i$ should also prefer the GRIM strategy. In summary, in this game scenario defective strategies are to be favored by both players.

Game scenario II, with slightly different QoS requirements of $(\Theta^i_{req}, \Delta^i_{req}, \Xi^i_{req}) = (0.1, 0.05, 0.02)$ and $(\Theta^{-i}_{req}, \Delta^{-i}_{req}, \Xi^{-i}_{req}) = (0.4, 0.031, 0.02)$, is also summarized in Figure 5.32 with the graphs marked in squares and diamonds. The MSG QoS outcomes as shown in

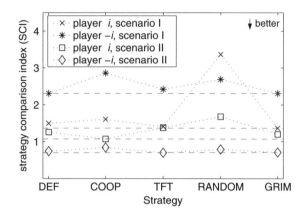

Figure 5.32 Strategy comparison of the success for QoS support with the help of a Strategy Comparison Index (SCI) as summarizing value for success in supporting QoS against the generality of all opponent strategies. Reproduced with permission from L. Berlemann, G. R. Hiertz, B. Walke, and S. Mangold, 'Strategies for Distributed QoS Support in Radio Spectrum Sharing,' in Proc. of IEEE International Conference on Communications, ICC 2005, Seoul, Korea, 16–20 May 2005. © 2005 IEEE

Figures 5.30 and 5.31 are not depicted for game scenario II. In this scenario, the strategy of COOP is the best for player i and the TFT strategy for player $-i$. As player i has a required throughput of 0.1, the competition for the medium is less severe, compared with the previous scenario, leading to an advantage in strategies with cooperative game outcomes of (C, C). Here, the best response optimizes the players' payoff with less destructive interference, such as blocking of the medium, for the opponent. Player $-i$ with $\Theta_{req}^{-i} = 0.4$ is more sensitive to the opponent's behavior and has a better QoS with the TFT trigger strategy.

5.5.14 Game Approach as Policy*

The application of game theory in spectrum sharing scenarios enables a distributed coordination of multiple cognitive radios sharing the same spectrum opportunity. The above developed and evaluated strategies from our game approach are specified in this section in a machine-understandable policy description language, such that the cognitive radio is capable of reasoning about spectrum usage. The XML based DARPA XG policy language from Section 5.4 is hence taken as basis. The identification of a spectrum opportunity is to be done by applying additional policies and is not part of the sharing algorithms from the game theory-based approach.

This section describes, as an example, how spectrum sharing algorithms in general can be transferred from research and development on an operating cognitive radio. The expressiveness of the policy framework thereby determines the complexity and capability of

*L. Berlemann, S. Mangold, and B. H. Walke, "Policy-based Reasoning for Spectrum Sharing in Cognitive Radio Networks," in Proc. of 1st IEEE International Symposium on New Frontiers in Dynamic Spectrum Access Networks, DySPAN2005, Baltimore MD, USA, 8–11 November 2005.

realizing the sharing policies on the cognitive radio. These define the operational parameters for the underlying radio access technology of the cognitive radio.

5.5.14.1 Mapping Games to a Policy

The transfer of the game theory notation is initiated in defining device capabilities, game parameters and the behavior of a player in Policy 5.2. The behavior, as handpicked allocation of the radio resource, is specified as usage description (useDesc). Thereafter the opponent's behavior is classified in order to characterize the spectrum opportunity for the next stage. In taking the own behavior of the present stage into account, every permutation of the players' behavior (here in total four) leads to a dedicated opportunity description (OppDesc) as demonstrated in Policy 5.3.

> **Policy 5.2** Game parameters and the behaviors of a players expressed in shorthand notation of the DARPA XG policy language. Reproduced with permission from L. Berlemann, S. Mangold, and B. H. Walke, 'Policy-based Reasoning for Spectrum Sharing in Cognitive Radio Networks,' in Proc. of 1st IEEE International Symposium on New Frontiers in Dynamic Spectrum Access Networks, DySPAN2005, Baltimore MD, USA, 8–11 November 2005. © 2005 IEEE

```
1   (DeviceCap (id GameTheoryProfile)
    (hasPolicyDefinedParams STAGEduration Theta_dem
    Delta_min Theta_req Delta_dem)
    (hasPolicyDefinedBehaviors ObserveStage ClassifyBehavior
    BestResponse))

2   (TimeDuration (id STAGEduration)
    (boundBy Device) (unit msec))

3   (Boolean (id OpponentCooperating))
    (Boolean (id SelfCooperating))

4   (UseDesc (id Defect) (xgx '(and
    (: = Theta_dem BestResponse, (oppAction))
    (: = Delta_dem BestResponse, (oppAction))
    (: = SelfCooperating BoolFalse)))

5   (UseDesc (id Cooperate) (xgx '( and
    (: = Theta_dem Theta_req)
    (: = Delta_dem Delta_min)
    (: = SelfCooperating BoolTrue)))
```

Simple static strategies can be defined as PolicyRule, as is demonstrated in Policy 5.4. Complex strategies, which take the behavior of the opponent into account, are realized as a group PolicyGrp of policy rules PolicyRule. Thereby, each state transition of the strategies' state machine is reflected by a policy rule, defining the reaction of a player to the opponent's behavior in taking own behavior into account. This is illustrated by comparing two dynamic trigger strategies specified in Policy 5.5 and 5.6 and depicted in Figures 5.34 and 5.35.

Policy 5.3   The classification of the opponent's behavior expressed in short-hand notation of the DARPA XG policy language. Reproduced with permission from L. Berlemann, S. Mangold, and B. H. Walke, 'Policy-based Reasoning for Spectrum Sharing in Cognitive Radio Networks,' in Proc. of 1st IEEE International Symposium on New Frontiers in Dynamic Spectrum Access Networks, DySPAN2005, Baltimore MD, USA, 8–11 November 2005. © 2005 IEEE

```
1   (OppDesc (id OwnCoop_OpponentCoop)
    (xgx '(and
    (invoke (within STAGE) ObserveStage
    ObsParam Observation.ownQoS
    ObsParam Observation.oppQoS)
    (invoke (at-end-of STAGE)
    ClassifyBehavior Observation.ownQoS
    Observation.oppQoS
    OpponentCoop OpponentCooperating)
    (and (eq OpponentCooperating BoolTrue)
    (eq SelfCooperating BoolTrue)'))

2   (OppDesc (id OwnDef_OpponentCoop)
    (xgx '(and
    (invoke (within STAGE) ObserveStage
    ObsParam Observation.ownQoS
    ObsParam Observation.oppQoS)
    (invoke (at-end-of STAGE)
    ClassifyBehavior Observation.ownQoS
    Observation.oppQoS
    OpponentCoop OpponentCooperating)
    (and (eq OpponentCooperating BoolTrue)
    (eq SelfCooperating BoolFalse))'))

3   (OppDesc (id OwnCoop_OpponentDef)
    (xgx '(and (invoke (within STAGE) ObserveStage
    ObsParam Observation.ownQoS
    ObsParam Observation.oppQoS)
    (invoke (at-end-of STAGE)
    ClassifyBehavior Observation.ownQoS
    Observation.oppQoS
    OppCoop OpponentCooperating)
    (and (eq OpponentCooperating BoolFalse)
    (eq SelfCooperating BoolTrue))'))

4   (OppDesc (id OwnDef_OpponentDef)
    (xgx '(and
    (invoke (within STAGE) ObserveStage
    ObsParam Observation.ownQoS
    ObsParam Observation.oppQoS)
    (invoke (at-end-of STAGE)
    ClassifyBehavior Observation.ownQoS
    Observation.oppQoS OppCoop OpponentCooperating)
    (and (eq OpponentCooperating BoolFalse)
    (eq SelfCooperating BoolFalse))'))
```

Policy 5.4 COOP and DEF strategy expressed in shorthand notation of the DARPA XG policy language. Reproduced with permission from L. Berlemann, S. Mangold, and B. H. Walke, 'Policy-based Reasoning for Spectrum Sharing in Cognitive Radio Networks,' in Proc. of 1st IEEE International Symposium on New Frontiers in Dynamic Spectrum Access Networks, DySPAN2005, Baltimore MD, USA, 8–11 November 2005. © 2005 IEEE

1	(PolicyRule (id StrategyCOOP) (selDesc S1) (deny FALSE) (oppDesc AnyOpp) (useDesc Cooperate))
2	(PolicyRule (id StrategyDEF) (selDesc S1) (deny FALSE) (oppDesc AnyOpp) (useDesc Defect))

Policy 5.5 TFT strategy expressed in shorthand notation of the DARPA XG policy language. Reproduced with permission from L. Berlemann, S. Mangold, and B. H. Walke, 'Policy-based Reasoning for Spectrum Sharing in Cognitive Radio Networks,' in Proc. of 1st IEEE International Symposium on New Frontiers in Dynamic Spectrum Access Networks, DySPAN2005, Baltimore MD, USA, 8–11 November 2005. © 2005 IEEE

1	(PolicyGrp (id StrategyTit-For-Tat) (equalPrecedence TRUE) (polMembers TFTCoop1 TFTCoop2 TFTDefect1 TFTDefect2))
2	(PolicyRule (id TFTCoop1) (selDesc S1) (deny FALSE) (oppDesc OwnCoop_OpponentCoop) (useDesc Cooperate))
3	(PolicyRule (id TFTDefect1) (selDesc S1) (deny FALSE) (oppDesc OwnCoop_OpponentDef) (useDesc Defect))
4	(PolicyRule (id TFTCoop2) (selDesc S1) (deny FALSE) (oppDesc OwnDef_OpponentCoop) (useDesc Cooperate))
5	(PolicyRule (id TFTDefect2) (selDesc S1) (deny FALSE) (oppDesc OwnDef_OpponentDef) (useDesc Defect))

5.5.14.2 Behaviors as Policies

Policy 5.2 defines device capabilities and describes game parameters and the behavior of a player:

- Line 1 – The capability description of parameters and processes that a cognitive radio has to provide in order to apply game theory-based policies. They are used in the following policies.
- Line 2 – The duration of an SSG is provided by the cognitive radio, typically it has a duration of 100 ms.

- Line 3 – Parameters to indicate if an opponent is cooperating and for storing the player's own behavior for the present stage.
- Line 4 – The behavior of defection as usage description resulting in a concrete action. Best response to the expected opponent's action oppAction to optimize the own utility defined in the process BestResponse. This process is not defined here.
- Line 5 – The behavior of cooperation as usage description resulting in a concrete action, i.e. reduction of the period length Δ_{dem} to Δ_{min} and demanding the required throughput $\Theta_{dem} = \Theta_{req}$. Note that these parameters specify a dedicated allocation pattern for one stage and are to be provided by the cognitive radio similar to MaxTransmitPower in Policy 5.1, line 8.

Policy 5.6   GRIM strategy expressed in shorthand notation of the DARPA XG policy language. Reproduced with permission from L. Berlemann, S. Mangold, and B. H. Walke, 'Policy-based Reasoning for Spectrum Sharing in Cognitive Radio Networks,' in Proc. of 1st IEEE International Symposium on New Frontiers in Dynamic Spectrum Access Networks, DySPAN2005, Baltimore MD, USA, 8–11 November 2005. © 2005 IEEE

1	(PolicyGrp (id StrategyGRIM) (equalPrecedence TRUE) (polMembers GRIMCoop1 GRIMDefect1 GRIMDefect2 GRIMDefect3))
2	(PolicyRule (id GRIMCoop1) (selDesc S1) (deny FALSE) (oppDesc OwnCoop_OpponentCoop) (useDesc Cooperate))
3	(PolicyRule (id GRIMDefect1) (selDesc S1) (deny FALSE) (oppDesc OwnCoop_OpponentDef) (useDesc Defect))
4	(PolicyRule (id GRIMDefect2) (selDesc S1) (deny FALSE) (oppDesc OwnDef_OpponentCoop) (useDesc Defect))
5	(PolicyRule (id GRIMDefect3) (selDesc S1) (deny FALSE) (oppDesc OwnDef_OpponentDef) (useDesc Defect))

Policy 5.3 introduces the classification of the opponent's behavior.

- Line 1 – This OwnCoop_OpponentCoop opportunity description has three tests: (i) The process ObserveStage observes all allocations during a stage and has the observed QoS of a player and of its opponent as output parameter. (ii) The process ClassifyBehavior, invoked at the end of a stage, determines the players' QoS of the last stage in observing spectrum usage. The process decides about the opponent's behavior contained as output in the Boolean variable OpponentCooperating. (iii) The opponent cooperates (OpponentCooperating = TRUE) and player self cooperates in the considered stage (SelfCooperating = TRUE). In case all these tests are met the player concludes that

both players were cooperating and regards the spectrum opportunity as OwnCoop_
OpponentCoop

- Line 2 – The OwnDef_OpponentCoop opportunity description is similar to the Own-
  Coop_OpponentCoop description, as well as the last of the three tests. The player self is
  defecting in the considered stage SelfCooperating = FALSE. In the case of all tests being
  met, the player assumes own defection while the opponent was cooperating
- Line 3 – This OwnCoop_OpponentDef opportunity description has three tests. (i) The
  process ObserveStage observes during a stage all allocations and has the observed QoS
  of a player and of its opponent as output parameter. (ii) The process ClassifyBehavior,
  invoked at the end of a stage, determines the players' QoS of the last stage in observing spectrum usage. The process decides about the opponent's behavior contained as
  output in the Boolean variable OpponentCooperating. (iii) The opponent defects if
  OpponentCooperating = FALSE and player self cooperates in the considered stage when
  SelfCooperating = TRUE. In the case of all three tests being met, the player concludes that
  it was cooperating while the opponent was defecting.
- Line 4 – The OwnDef_OpponentDef opportunity description is similar to the Own-
  Coop_OpponentDef description, besides the last of the three tests. The player self was
  defecting in the considered stage SelfCooperating = FALSE. In the case of all tests being
  met, the player assumes that both players were defecting.

5.5.14.3 Static Strategies as Policies

Static strategies are the continuous application of one behavior without regarding the opponent's strategy. In static strategies, the state model contains one single state. In the approach
under consideration here, the set of available static strategies is reduced to two; the cooperation strategy (COOP) is characterized by cooperating always, independently of the opponent's
influence on the player's utility. The COOP strategy is to the benefit of a player if the opponent
cooperates as well. Figure 5.33(a) illustrates this simple strategy of following a cooperative
(C) behavior, as specified in line 5 of Policy 5.2.

Equivalently to the COOP strategy, the defection strategy (DEF) consists of a permanently chosen behavior of defection (D). Figure 5.33(b) illustrates the DEF strategy as a

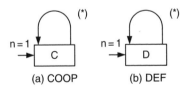

Figure 5.33 The static strategies of permanent cooperation (a) and defection (b) as state machines.
Reproduced with permission from L. Berlemann, S. Mangold, and B. H. Walke, 'Policy-based Reasoning for Spectrum Sharing in Cognitive Radio Networks,' in Proc. of 1st IEEE International Symposium
on New Frontiers in Dynamic Spectrum Access Networks, DySPAN2005, Baltimore MD, USA,
8–11 November 2005. © 2005 IEEE

state machine. The static strategies of permanent cooperation and defection are expressed in Policy 5.4 with the following meaning:

- Line 1 – The strategy COOP realized as a PolicyRule for the selector description S1 defined in Policy 5.1. Independently of the opponent behavior, i.e., for any spectrum opportunity AnyOpp, the player cooperates. This is specified by the usage description Cooperate
- Line 2 – The strategy DEF realized as a PolicyRule. Independent of the opponent behavior, i.e., for any spectrum opportunity AnyOpp, the player cooperates. This is specified by the usage description Defect

5.5.14.4 Dynamic Trigger Strategies as Policies

A trigger strategy is a dynamic strategy where the transition from one state to another state is event-driven (Osborne and Rubinstein, 1994); an observed event triggers a behavior change of a player. Depending on the number of states (the number of behaviors a player may select), a large number of trigger strategies is possible. For the sake of simplicity, the familiar Grim (GRIM) and Tit-For-Tat (TFT) trigger strategies are applied in the following. A player with a GRIM strategy punishes the opponent for a single deviation from cooperation with a defection forever.

The initial state of the GRIM strategy, selected at the first stage of the MSG is, however, the cooperation. The player cooperates as long as the opponent cooperates, and the transition to defection is triggered by the opponent's defection. See Figure 5.35 for an illustration of the state machine of the GRIM strategy and its stepwise transfer to a policy. The TFT strategy selects cooperation as long as the opponent is cooperating, similarly to the GRIM strategy, and also with cooperation in the initial stage. An opponent's defection in stage N triggers a state transition and is punished by defection in the following stage $(N + 1)$, as illustrated in Figure 5.34. However, in contrast to the GRIM strategy, TFT changes back to cooperative behavior as soon as the opponent is cooperating again.

Policy 5.6 expresses the GRIM strategy from the game theory-based approach. The corresponding state machine is depicted in Figure 5.35, which contains references to the

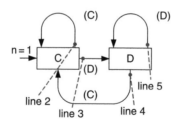

Figure 5.34 The trigger strategy Tit-For-Tat, specified in Policy 5.5 as state machine. Reproduced with permission from L. Berlemann, S. Mangold, and B. H. Walke, 'Policy-based Reasoning for Spectrum Sharing in Cognitive Radio Networks,' in Proc. of 1st IEEE International Symposium on New Frontiers in Dynamic Spectrum Access Networks, DySPAN2005, Baltimore MD, USA, 8–11 November 2005. © 2005 IEEE

corresponding line of the description in the policy language. In detail, the lines of Policy 5.6 have the following meaning:

- Line 1 – The strategy GRIM consists of four policy rules: GRIMCoop1 (line 2), GRIMDefect1 (line 3), GRIMDefect2 (line 4) and GRIMDefect3 (line 5). All policies in the group have the same priority as indicated by the property equalPrecedence.
- Line 2 – The policy rule GRIMCoop1 for operation matching selector S1 (defined above). In the case of a cooperating opponent and own cooperation, i.e., the opportunity is regarded as OwnCoop_OpponentCoop (Policy 5.3, line 1), the player chooses the behavior of cooperation by following the usage description Cooperate (Policy 5.2, line 4). The term Deny = FALSE indicates that the rule represents a valid opportunity.
- Line 3 – The policy rule GRIMDefect1. In the case of own cooperation and a defecting opponent, i.e., the opportunity is regarded as OwnCoop_OpponentDef (Policy 5.3, line 2), the player defects following the usage description Defect (Policy 5.2, line 3).
- Line 4 – The policy rule GRIMDefect2 In the case of own defection and a cooperating opponent (Policy 5.3, line 3), the player defects following the usage description Defect (Policy 5.2, line 3).
- Line 5 – The policy rule GRIMDefect3. In the case of own defection and a defecting opponent (Policy 5.3, line 2), the player defects following the usage description Defect (Policy 5.2, line 3).

The TFT strategy, as illustrated in Figure 5.34, is specified in Policy 5.5. The description is analogous to that of the GRIM strategy, reflecting the similarity of the respective state machines:

- Line 1 – The strategy Tit-For-Tat consists of four policy rules: TFTCoop1 (line 2), TFTDefect1 (line 3), TFTCoop2 (line 4) and TFTDefect2 (line 5). The property equalPrecedence indicates that all policies in the group have the same priority.
- Lines 2, 3 and 5 are the same as in Policy 5.6 of the GRIM strategy.
- Line 4 – The reaction on a cooperating opponent in the case of own defection (Policy 5.3, line 3) is different. This dissimilarity is marked bold in the policies of GRIM and TFT.

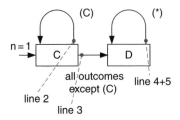

Figure 5.35 The trigger strategy GRIM, specified in Policy 5.6 as state machine. Reproduced with permission from L. Berlemann, S. Mangold, and B. H. Walke, 'Policy-based Reasoning for Spectrum Sharing in Cognitive Radio Networks,' in Proc. of 1st IEEE International Symposium on New Frontiers in Dynamic Spectrum Access Networks, DySPAN2005, Baltimore MD, USA, 8–11 November 2005. © 2005 IEEE

Here, the player cooperates following the usage description Cooperate (Policy 5.2, line 4), reflecting the different state transitions in Figures 5.35 and 5.34.

5.5.15 Learning from Spectrum Sharing Games

Support of QoS in wireless networks that share the same spectrum is a problem in the decentralized coordination of medium access. The application of games allows an aimed interaction for mutual coordination and provides an analysis of the competition for the utilization of a shared radio spectrum. The analysis and simulation results indicate that cooperation is an achievable equilibrium that may improve the overall efficiency in spectrum sharing. Traffic requirements that are imposed by services and applications determine whether the selected strategies should pursue cooperation, or ignore other radio systems, thus leading to games of defection. In defective scenarios, where cooperation is not achievable, a regulating intervention, for example the specification of some MAC parameters, may be advantageous. However, it is the intention of this section to enable competing wireless networks to guarantee QoS support with minimum regulation. Game models of multiple players and learning in games how to facilitate overcoming of insufficient information about opponents are the next steps in further improving the reduction of mutual interference.

This section's game approach has one essential requirement, namely the need for a predictable allocation of spectrum, for the purpose of enabling wireless networks to achieve interaction. The resulting periodic spectrum allocations may introduce artificial queuing times, and hence require additional buffering, segmentation or even transmissions of frames that do not carry information.

The specification examples of spectrum sharing strategies as policies illustrate the general applicability of the policy framework: It offers a common method for translating strategies represented as state machines to the DARPA XG policy language.

6

Proposed Enablers for Realizing Vertical Spectrum Sharing

Today's framework for radio spectrum regulation and spectrum usage is undergoing fundamental changes. In the face of scarce radio resources, regulators, industry, and the research community are initiating promising approaches towards a more flexible spectrum usage such as, for example, cognitive radios for dynamic spectrum assignment. In this chapter, medium access control protocols for cognitive radios that operate as secondary devices in parts of the spectrum originally licensed to other primary radio services are discussed. They identify free spectrum, coordinate its usage and release it again when this is required by the licensed radio systems. This chapter concentrates on several enablers for overlay vertical spectrum sharing. Underlay vertical spectrum sharing based on UWB technologies is not discussed here but can for instance be found in the context of high data rate wireless personal area networks in Chapter 6 of Walke *et al.* (2006).

This chapter is structured as follows. A low cost alternative to UMTS/3G is introduced in Section 6.1 with 802.11 WLANs operating in a Frequency Division Duplex (FDD) mode. This introduces new business opportunities to operators with an (locally) unused FDD license, as FDD WLANs might be operated as secondary systems in this spectrum under the control and billing of the license holder. The reuse of TV bands with an operator-assisted cognitive radio approach, based on dual beaconing enhancements in the example of WLAN, is discussed in Section 6.2. Spectrum load smoothing derived from the 'waterfilling' principle known from information theory is introduced in Section 6.3. It is applied in 802.11 in two variants, namely, with and without the usage of reservations.

6.1 Frequency Division Duplex for Wi-Fi: FDD WLANs

We learned in Section 5.1 that IEEE 802.11 Wireless LAN can be seen as a low cost communication system well suited to providing spectrum access for many of today's communication services. It is nearly ideally suited to support data communication with moderate quality-of-service requirements, particularly when operating in unlicensed spectrum. Since licensed spectrum will be regulated in a way referred to as technology neutral in the future, it is only natural to use WLAN in licensed spectrum as well. However, much of today's spectrum for next generation cellular networks will be marked as paired spectrum for Frequency Division Duplex (FDD) with dedicated uplink and downlink spectrum. One reason for this preference for FDD is that so-called cross-operator interference can be more easily coordinated with paired spectrum. For example, at national borders, FDD simplifies the coordination between the different regulatory bodies.

Today's WLAN, with its listen-before-talk protocol where communicating devices operate in the same spectrum, is therefore not an obvious candidate technology to be used in paired spectrum. This could be easily solved by extending the existing IEEE 802.11 WLAN standard towards FDD support. Figure 6.1 illustrates how paired FDD spectrum access could be realized for WLAN IEEE 802.11. We refer to this as the important enabler 'FDD WLAN' for vertical spectrum sharing.

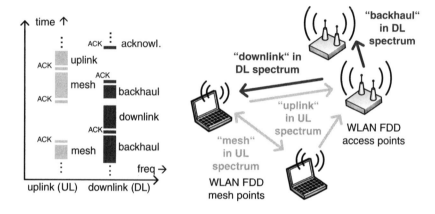

Figure 6.1 Wireless LAN operating in mesh mode in a paired spectrum. Paired spectrum is traditionally used for frequency division duplex of 3G and beyond-3G systems with a balanced uplink versus downlink capacity ratio. Traditional communication, such as circuit-switched voice calls, require the same channel capacity in downlink and uplink. With WLAN FDD operating in mesh mode, the contention protocol of IEEE 802.11 is modified to operate at the two channels, instead of only one. Some devices then play the traditional role of a base station and will transmit on downlink spectrum. Other devices play the role of mobile devices and will transmit on the uplink spectrum back to the base station, or in direct mode directly to each other

Paired spectrum is traditionally used for frequency division duplex of 3G and beyond-3G systems with a balanced uplink versus downlink capacity ratio. Traditional communication, such as circuit-switched voice calls, require the same channel capacity in downlink and uplink. With WLAN FDD operating in mesh mode, the contention protocol of IEEE 802.11 must be modified to operate on two channels instead of one. Some devices (for example, an 802.11 access point) then play the traditional role of a base station and transmit on downlink spectrum. Other devices (the 802.11 stations and mesh points) play the role of a traditional mobile device and transmit on the uplink spectrum back to the base station. In total, all devices would receive and transmit on different channels and a WLAN FDD system is realized.

Future WLAN and mesh networks support direct communication between stations. In direct communication, stations communicate with each other by using the uplink spectrum.

Finally, as indicated in the figure, mesh points of future WLAN mesh networks may backhaul over the downlink spectrum.

Why would an operator want to deploy FDD WLAN networks? The answer is straightforward: there are standards that fail commercially, or that experience long market take-up times. For example, 3G with its many promises for universal communication, did not meet customer demands for a long time. Only recently did the apparent benefits of 3G technology compared with earlier standards start to convince customers. As result, for a long time the allocated 3G spectrum was not used at all. Our spectrum measurements, which we performed in 2006, demonstrated the limited usage of 3G networks at that time (see Section 2.3.4 of this book).

Another example is IEEE 802.16 (WiMAX). This is a complicated standard and the technology is expensive. It often requires paired spectrum. Throughout recent years, in the hope of creating competition between new and incumbent telecom operators, WiMAX has been supported by regulators, and paired spectrum has been licensed in many countries. However, today's commercial implementations of WiMAX are TDD systems. Until today there were many countries and regions where TDD WiMAX had not yet been deployed and FDD WiMAX was not commercially available. We should again expect a long take up time for both technologies.

The cases of 3G and WiMAX illustrate that a cost-effective solution may be helpful. A low cost alternative to 3G and WiMAX, at least during the early years of deployment, could be FDD WLAN. When the new and desired technology is rolled out, the FDD WLAN could gradually be replaced by the desired technology by sharing the paired spectrum vertically. This introduces new business opportunities to operators with (locally) unused FDD licenses, because FDD WLANs might be operated as secondary systems in this spectrum, under the control and the billing of the license holder.

6.2 Operator Assisted Cognitive Radio with Beaconing*

Operator assistance (Mangold *et al.*, 2006b, c) refers to the combination of centralized means for coordination with the decentralized paradigm of cognitive radio, as outlined in Section 3.3.3. The advantage of centralized approaches is that they permit regulators to remain in control of spectrum usage, and allow them to direct how the spectrum is used. This is of particular interest in vertical sharing. On the other hand, the advantage of decentralized approaches is their great flexibility, which is especially helpful in horizontal sharing. To protect licensed radio systems in vertical sharing scenarios, while gaining from the flexibility of decentralized approaches, radio systems such as an established cellular network may assist cognitive radios in identifying underutilized spectrum. One approach towards this centralized operator assistance, namely beaconing, is introduced and initially evaluated in the following. Operator assistance refers to the combination of centralized means for coordination with the decentralized paradigm of cognitive radio, as outlined in Section 3.3.3.

---

* S. Mangold, A. Jarosch, and C. Monney (2006) Operator Assisted Dynamic Spectrum Assignment with Dual Beacons. IEEE International Zurich Seminar on Communications IZS 2006. Zurich, Switzerland, 22–24 February 2006. © 2006 IEEE

6.2.1 Existing Standard Beaconing Concepts

Two main concepts for such a signaling mechanism are usually discussed in the literature (FCC, 2003c; Hulpert, 2005). Both proposals are based on beaconing mechanisms for
broadcasting either permissions (grants) for cognitive radios to access spectrum, or alternatively, denials of spectrum access for cognitive radios. See Figure 6.2 for an illustration of the
denial and grant beaconing concepts. Both concepts have advantages and disadvantages, and
are known for their limited reliability. The two major limitations of the two approaches are as
described below.

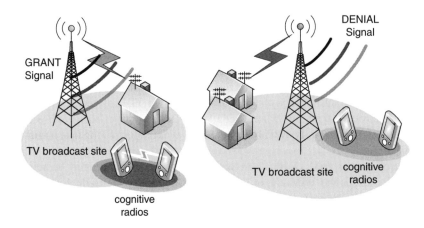

Figure 6.2   Two existing state-of-the-art beaconing concepts. Left: cognitive radios transmit only if
and only if a grant beacon is detected. Right: cognitive radios transmit only if and only if no denial
beacon is detected

If busy (denial) beacons are applied, it may occur that a cognitive radio that did not receive
this beacon will access the spectrum anyway, because of shadowing or the hidden station
problem. Although spectrum access was denied by the primary radio system, a cognitive radio,
even in close vicinity, may utilize the spectrum because of the missed detection of the beacon.
Therefore the concept is not reliable.

If permission (grant) beacons are applied, another problem occurs. Cognitive radios could
utilize the spectrum because they detected the permission from a primary radio system. However, another primary radio system that could be hidden from the first primary may require the
spectrum, but there is no mechanism to prevent the cognitive radio from utilizing it. Again, a
reliability problem exists.

6.2.2 What is a Beacon?

We briefly summarize, in the following, initial thoughts on how a signaling mechanism could
be implemented.

There are various approaches for beacon implementation. The simplest approach would
be the out-of-band emission of a single harmonic electromagnetic wave, modulated only by
on/off switching. This wave would be associated with a predefined band or frequency channel,
to signal this channel's availability. Instead of out-of-band emission, the harmonic wave could

be realized by in-band transmission during times when the respective band is unused. Beacons can also be implemented as frames with preambles and information fields. Finally, beacons could be realized as underlay low-power signals similar to ultrawideband.

Independent of the implementation of beacons, what is common to all discussed approaches is that the beacons are transmitted at some power, and detected at some sensitivity. Beacons may or may not reach the intended cognitive radios, depending on the shadowing/fading situation. Our proposed dual beacon approach is independent of the way a beaconing is implemented.

6.2.3 Improved Signaling Mechanism with Dual Beacons

An improved signaling mechanism to coordinate spectrum usage by cognitive radios is as simple as it is efficient. It is proposed that cognitive radios be coordinated with two, instead of one, single signal. One signal would be similar to a denial beacon, and one signal would be similar to a grant beacon.

Basically, cognitive radios refrain from utilizing the spectrum if no grant beacon is detected. As a result, as long as a primary radio system is not transmitting a grant beacon, there will be no source of interference from cognitive radios. Cognitive radios utilize the spectrum if, and only if, a grant beacon is detected, and no denial beacon is detected. This simple logic ensures higher reliability by allowing the necessary flexibility in coordinating the reuse of licensed radio spectrum. If a grant beacon and a denial beacon are detected at the same time, a cognitive radio refrains from spectrum utilization. Therefore, in hidden station scenarios, a primary can always ensure that its environment is free of interference, and still coordinate the secondary, i.e. the spectrum utilization by cognitive radios.

The presented signaling mechanism is a step towards operator assistance for dynamic spectrum assignment.

6.2.4 Beacon Implementation in IEEE 802.11

This section introduces implementation of the dual beaconing approach in 802.11. A periodically transmitted 802.11 management frame, namely the beacon frame, is extended here to realize our approach. At the end of the 802.11 beacon, an individual flag for each of the 69 TV channels is added, see Table 6.1. This flag represents either the grant beacon '1' or the denial beacon '0' for allowing secondary access to the corresponding frequency band of the TV channel. Such a proprietary or vendor specific extension of the standard 802.11 beacon is supported by legacy 802.11 stations, which ignore all addition information elements added at the end of the legacy beacon. The regulatory domain, i.e. the country in which the beacon is broadcast, is already part of the beacon frame (added by IEEE 802.11i). This regulatory domain determines the concrete mapping of TV channel number to frequency band. For this

Table 6.1 Extension to standard 802.11 beacon frame format for enabling reuse of TV channels with dual beaconing

Order	Information	Notes
33	Denial/grant flag for each TV channel	69 TV channels, size: 69 bit

mapping, a data base containing the country specific frequency bands needs to be pre-installed on the cognitive radio device.

The policy language introduced in Section 5.4 also offers the possibility of specifying these characteristics of the radio spectrum, as shown below in Section 6.2.6.

6.2.5 Evaluation

Figure 6.3 illustrates the evaluation scenario. Let one station S communicate with one base station BS1. One cognitive radio is located at an arbitrary position. Let BS1 continuously broadcast a denial beacon to protect its communication with station S against harmful interference from cognitive radios, and let the neighboring base station BS2 continuously broadcast a grant beacon. For the evaluation, we assume that the station S moves on a direct path from BS1 to BS2. The cell radius is assumed to be $r = 300\ m$. The base stations BS1 and BS2 are located at $(x, y) = (-300\ m, 0\ m)$ and $(x, y) = (+300\ m, 0\ m)$, respectively. Hence, station S moves from $-300\ m$ to the cell edge at $0\ m$ and into the neighboring cell area up to $+300\ m$.

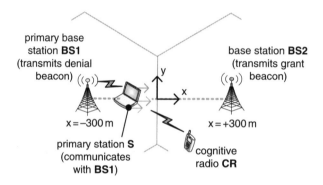

Figure 6.3 Evaluation scenario. One station S communicates with base station BS1. One cognitive radio located at an arbitrary position may create unacceptable interference. BS1, in addition, broadcasts a denial beacon and the neighboring base station BS2 broadcasts a grant beacon. Reproduced with permission from S. Mangold, A. Jarosch, and C. Monney (2006) Operator Assisted Dynamic Spectrum Assignment with Dual Beacons. IEEE International Zurich Seminar on Communications IZS 2006. Zurich, Switzerland, Feb. 22–24, 2006. © 2006 IEEE

We use the Monte-Carlo approach for our simulation to average over a reasonable set of possible positions for the cognitive radio device cognitive radio. In the simulation, the station S that communicates with base station BS1, moves from BS1 to BS2, which is equivalent to a course of 600 m. Depending on transmission powers, sensitivity levels, and locations of the devices, the cognitive radio may or may not impose significant interference to the reception of station S. See Table 6.2 for the selected values of the model parameters. It is noted that the results change significantly when changing these numbers, but remain qualitatively the same. In the following, four different scenarios are evaluated.

① **State of the art**: The cognitive radio follows a grant beacon only; the grant beacon is transmitted by BS2. The corresponding logic is shown in Equation (6.1). Evaluation results are shown in Figure 6.4(a).

Table 6.2 Parameters of the evaluation model used for assessing beaconing concepts. Reproduced with permission from S. Mangold, A. Jarosch, and C. Monney (2006) Operator Assisted Dynamic Spectrum Assignment with Dual Beacons. IEEE International Zurich Seminar on Communications IZS 2006. Zurich, Switzerland, Feb. 22–24, 2006. © 2006 IEEE

Parameter	Value
Transmission power of BS1 to station S	1000 mW
Beacon transmission power of BS1 and BS2	500 mW
Denial beacon transmission power of station S in scenario ④	50 mW
Beacon detection sensitivity at cognitive radio for all beacons	–72 dBm
System noise	–90 dBm
Noise figure of all devices	10 dB
Background noise not including devices' noise figures	–100 dBm
Transmission power of cognitive radio	10 mW
System frequency	5.2 GHz
Channel bandwidth	20.0 MHz

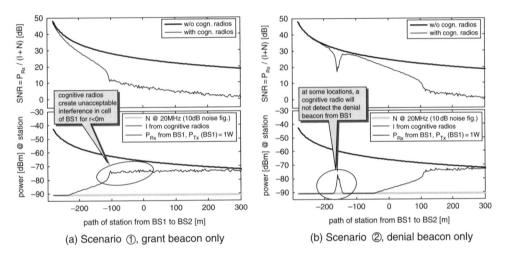

(a) Scenario ①, grant beacon only (b) Scenario ②, denial beacon only

Figure 6.4 Evaluation results. The resulting powers and interferences in dBm are shown in the bottom section of each figure, and the corresponding SNRs are shown in the top section. In scenario ①, with decreasing distance between S and BS2, station S moves into an area where interference from the cognitive radio is more likely to occur. This interference remains constant once S has entered the cell of BS2. In scenario ②, some locations for the cognitive radio are assumed to be hidden to BS1 and therefore a cognitive radio located in this area may create significant interference to station S, even if S is located inside the cell of BS1. Reproduced with permission from S. Mangold, A. Jarosch, and C. Monney (2006) Operator Assisted Dynamic Spectrum Assignment with Dual Beacons. IEEE International Zurich Seminar on Communications IZS 2006. Zurich, Switzerland, Feb. 22–24, 2006. © 2006 IEEE

$$
\begin{aligned}
&\texttt{if } (GrantBeacon > -72\,dBm) \\
&\quad \{\texttt{access}\} \\
&\texttt{else} \\
&\quad \{\texttt{refrain}\}
\end{aligned}
\tag{6.1}
$$

② **State of the art**: The cognitive radio follows a denial beacon only; the grant beacon is transmitted by BS1. The corresponding logic is shown in Equation (6.2). Evaluation results are shown in Figure 6.4(b).

$$
\begin{aligned}
&\texttt{if (} DenialBeacon > -72\,dBm \texttt{)} \\
&\qquad \texttt{\{refrain\}} \\
&\texttt{else} \\
&\qquad \texttt{\{access\}}
\end{aligned}
\tag{6.2}
$$

③ **Novel approach**: The cognitive radio follows a grant and a denial beacon, the grant beacon is transmitted by BS2, and the denial beacon is transmitted by BS1. The corresponding logic is shown in Equation (6.3). Evaluation results are shown in Figure 6.5(a).

$$
\begin{aligned}
&\texttt{if (} GrantBeacon > -72\,dBm \texttt{)} \\
&\qquad + not\,(DenialBeacon > -72\,dBm \texttt{)} \\
&\qquad \texttt{\{access\}} \\
&\texttt{else} \\
&\qquad \texttt{\{refrain\}}
\end{aligned}
\tag{6.3}
$$

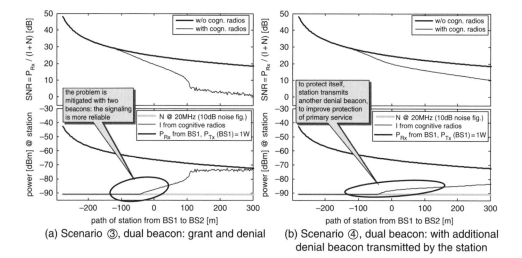

(a) Scenario ③, dual beacon: grant and denial   (b) Scenario ④, dual beacon: with additional denial beacon transmitted by the station

Figure 6.5 Evaluation results. The resulting powers and interferences in dBm are shown in the bottom part of each figure, and the corresponding SNRs are shown in the top. These curves are valid for the values as shown in Table 6.2. Scenario ③: The artificial hidden station problem exists as before, but a cognitive radio will not receive the grant beacon at this location, and will refrain from spectrum access. BS1 now protects its coverage area more reliably, and the combination of the grant and denial beacon concept ensures that station S will not observe interference within the cell of BS1. Scenario ④: The additional protection by the second denial beacon enables station S to maintain high SNR even while moving out of the coverage area of BS1, into the coverage area of BS2. Reproduced with permission from S. Mangold, A. Jarosch, and C. Monney (2006) Operator Assisted Dynamic Spectrum Assignment with Dual Beacons. IEEE International Zurich Seminar on Communications IZS 2006. Zurich, Switzerland, Feb. 22–24, 2006. © 2006 IEEE

④ *Novel approach*: The cognitive radio follows grant and denial beacons that are transmitted as in the previous scenario; in addition, station S transmits a second denial beacon for its own protection. The corresponding logic remains the same as before, and is indicated in Equation (6.3). Evaluation results are shown in Figure 6.5(b).

To calculate the received powers and interferences at specific locations, a simple free-space path loss model was selected as described in Equation (6.4).

$$
P_{Rx} = \begin{cases} P_{Tx} \cdot \dfrac{\lambda^2}{16\pi^2} \cdot \left(\dfrac{d_0}{d}\right)^2, d > d_0 \\ P_{Tx} \cdot \dfrac{\lambda^2}{16\pi^2}, d \leq d_0 \end{cases}, \text{ with } d_0 = 1m \tag{6.4}
$$

See Table 6.2 for the selected value for the system frequency, which translates to the wavelength λ. The selected metric to illustrate the performance of the different approaches is the SNR, which translates into channel capacity for the primary system by:

$$
C = B \cdot \log_2 \left(1 + \frac{P_{Rx}}{I + N}\right). \tag{6.5}
$$

We show numerical results for SNR $= P_{Rx}/(I + N)$ below.

6.2.5.1 Scenario ①: Grant Beacon

Figure 6.4(a) illustrates the evaluation results for the standard approach where only one grant beacon is used. The reader is reminded that the curves shown are only valid for the model parameter values that are given in Table 6.2. In the bottom figure, the resulting received power at station S, the observed interference, and the noise are indicated. In the top figure, the corresponding Signal-to-Noise-Ratio (SNR), which includes the interference, is shown. With decreasing distance between S and BS2, S moves into the area where interference from the cognitive radio is more likely, because any cognitive radio close to S will detect the grant beacon. The interference remains constant for $x > -100$ m because at any position close to the S, a cognitive radio will be detecting the grant beacon from BS2.

The problem of grant beacons is obvious: at the cell edge around $x = 0$ m, station S already observes interference from a cognitive radio. We conclude from this example that primary services cannot reliably be protected when using grant beacons only.

6.2.5.2 Scenario ②: Denial Beacon

A known proposal for solving the problem that was illustrated in the previous section is to use a denial beacon instead of a grant beacon. Evaluation results for this approach are shown in Figure 6.4(b). However, to illustrate the main problem of this denial beacon approach, we have created a hidden station scenario at a location around $x = -160$ m; cognitive radios that are close to this location are assumed to be hidden to BS1, which means that they cannot detect the denial beacon. Those cognitive radios may create significant interference to station S. The reader is again reminded that the curves shown depend on the selected values of the model parameters. It should be clear that the hidden station problem is the main reason why denial beacons are not a reliable solution for protecting primary services, even within the cell of BS1.

6.2.5.3 Scenario ③: Grant and Denial Beacon

Let us now discuss our proposed solution, to apply two beacons instead of one single beacon. The cognitive radio will access the spectrum if and only if the grant beacon is detected and no denial beacon is detected. The corresponding logical operation is shown in Equation (6.3). Evaluation results are shown in Figure 6.5(a). Note that in this figure, the artificial hidden station problem continues to exist, but due to the fact that at the respective locations a cognitive radio will not receive the grant beacon, it will refrain from spectrum access although no denial beacon was detected. This mitigates the hidden station problem. In addition, BS1 now is capable of protecting its coverage area more reliably and the combination of the grant and denial beacon concept now ensures that station S will not observe interference within the cell of BS1.

6.2.5.4 Scenario ④: Grant and Denial Beacon Plus Additional Denial Beacon from Station S

Finally, the protection of primary services can further be improved. It may occur that at some locations, the denial beacon is not detected due to shadowing or fading, but the grant beacon is detected. Then, interference from cognitive radio can be significantly high. To protect a primary station in such a scenario, we allow the station itself to transmit the denial beacon. This is evaluated in Figure 6.5(b). As expected, station S is able to reduce the interference even within the cell area of BS2 by simply transmitting an additional denial beacon at low power (here, 50 mW).

6.2.5.5 Lessons from the Evaluation

We introduced and evaluated a simple beaconing concept as a signaling mechanism for operator-assisted dynamic spectrum assignment. This novel and flexible way of coordinating spectrum sharing is a promising approach that aims for improvements in efficiency of spectrum usage. Such an improvement is clearly in the interest of regulators, industry, and the information society. It changes the way radio spectrum is regulated, and may help to make available the necessary radio resources that are required in the future. With the proposed beaconing concept, spectrum that would otherwise remain entirely unused, is made available for opportunistic usage. Spectrum usage is assisted in a reliable way with the help of a radio network that has clearly defined site locations and a trusted operator.

Operator assistance for cognitive radio systems is an innovative concept that would enable operators to provide a new type of service for spectrum owners (such as regulators or governments), or alternatively to the individual licensees. It may also be worth exploring our concept for coordinating unlicensed bands, mesh networks, and hot spot scenarios.

6.2.6 Dual Beaconing for the Reuse of TV Bands as Policy

The dual beaconing approach given above is specified in this section in the DARPA XG policy language. We assume that an enhanced 802.11 beacon is used for implementing the signaling of available frequencies for reuse and the spectrum access priorities. This beacon is here transmitted at 800 MHz.

The description framework is provided by DARPA XG and we specify here spectrum usage policies in the role of policy developers applying this framework. In Policy 6.1, groups of different candidate TV channels that might be reused for a limited time duration if allowed by the dual beaconing approach, are specified with the help of a selector description (SelDesc). Additionally, policy parameters are specified that are to be supported by (boundBy) the cognitive radio (cognitive radiodevice) to operate according to the specified policies. The radio transmission of a cognitive according to regulatory constrains is specified with the help of usage descriptions (UseDesc). Here, we show as an example how the transmission power can be limited. In Policy 6.2, the behavior of the cognitive radio for operation in TV bands is specified in four rules (PolicyRule).

The specification of the dual beaconing approach in a general policy language as presented here is independent of the underlying radio access technology.

Policy 6.1 Description of TV bands and a policy for operating in these bands expressed in shorthand notation of the DARPA XG policy language

policy	Description

(SelDesc (id Tvbands)
  (authDesc CH-OFCOM)
  (freqDesc Tvbands)
  (regnDesc CH)
  (timeDesc Until-123107)
  (devcDesc cognitive radiodevice))
	Selector description specifies all TV bands and devices allowed to operate in these bands. The issuing authority is the Swiss OFCOM (authDesc). The TV bands allowed for being re-used in general are described in the frequency band description (freqDesc). The policies validity is limited to Switzerland (regnDesc) and has a limited time duration.

(FreqDesc Tvbands
  (frequencyRanges Tvchannel52
  Tvchannel60 Tvchannel62))
	Three different TV channels (here 52, 60 and 62) may be reused: Their frequency ranges are grouped in this frequency band description Tvbands

(FrequencyRange (id Tvchannel52) (minValue 698) (maxValue 704) (unit MHz))	The frequency band (FrequencyRange) of Tvchannel52 is specified
(Frequency (id BeaconFreq) (magnitude 800.0) (unit MHz))	The frequency of the beacon is specified
(SignalEncoding (id GrandBeaconSig)) (SignalEncoding (id DenialBeaconSig))	The encoding of the grand and beacon signal
(Boolean(id GrandBeaconPresent)) (Boolean(id DenialBeaconPresent))	Parameters for indicating the presence of the beacons
(Power (id TransmitLimit) (magnitude 40.0) (unit mW))	A limit for the transmission power TransmitLimit is defined to 40 mW
(Power (id MaxTransmitPower) (boundBy Device))	MaxTransmitPower is declared and bound to a value provided by the protocol stack
(UseDesc (id UseTVbands) (xgx '(<= MaxTransmitPower TransmitLimit)'))	Usage description of limiting MaxTransmitPower to TransmitLimit. 'xgx' specifies an XG expression based on parameters to which the radio is able to provide values

Policy 6.2   Policy rules for reusing the TV bands with the dual beaconing approach expressed in shorthand notation of the DARPA XG policy language

policy	Explanation

(OppDesc
  (id GrandBeaconHeard)
  (xgx '(and (invoke SenseBeacon
  BeaconFreq BeaconFreq
  BeaconEnc GrandBeaconSig
  BeaconPresent
  GrandBeaconPresent))
  (eq GrandBeaconPresent
  BoolTrue))'))
	The opportunity condition that a grand beacon is present is determined in invoking the SenseBeacon process. The process must be supported by the cognitive radio and is not specified here. The beacon signal and beacon frequency are input parameters, the Boolean variable GrandBeaconPresent is the output parameter. In the case of the Grand Beacon being successfully received, the condition for characterizing the GrandBeaconHeard spectrum opportunity is satisfied

(OppDesc
  (id GrandBeaconNotHeard)
  (xgx '(and (invoke SenseBeacon
  BeaconFreq BeaconFreq
  BeaconEnc GrandBeaconSig
  BeaconPresent
  GrandBeaconPresent)
  (eq GrandBeaconPresent
  BoolFalse))'))
	The opportunity condition that a grand beacon is present is determined by invoking the SenseBeacon process. Again, the process must be supported by the cognitive radio and is not specified here. The output parameter variable GrandBeaconPresent is the output parameter. In the case of the Grand Beacon not being successfully received (output parameter variable GrandBeaconPresent is false), the condition for characterizing the GrandBeaconNotHeard spectrum opportunity is satisfied

(OppDesc
  (id DenailBeaconHeard)
  (xgx '(and (invoke SenseBeacon
  BeaconFreq BeaconFreq
  BeaconEnc DenailBeaconSig
  BeaconPresent
  DenailBeaconPresent))
  (eq DenailBeaconPresent
  BoolTrue))'))
	Similar to opportunity descriptions from above

(OppDesc
  (id DenailBeaconNotHeard)
  (xgx '(and (invoke SenseBeacon
  BeaconFreq BeaconFreq
  BeaconEnc DenailBeaconSig
  BeaconPresent
  DenailBeaconPresent)
  (eq DenailBeaconPresent
  BoolFalse))'))
	Similar to opportunity descriptions from above

6.3 Spectrum Load Smoothing*

Spectrum Load Smoothing (SLS) refers to the application of 'waterfilling', known from information theory, to the medium access of resource-sharing wireless networks. The objective of SLS is to enable the distributed Quality-of-Service (QoS) support in shared spectrum in an intelligent, cognitive, way. In using SLS, the competing wireless networks aim simultaneously at an equal, overall smoothed utilization of the spectrum. Thus, a decentralized coordinated, opportunistic usage of spectrum is implemented. In observing past usage of the radio resource, the wireless networks interact and redistribute their allocations of the spectrum under consideration of their individual QoS requirements. Due to the principle of SLS, these allocations are redistributed to less utilized or unallocated quantities of the transmission medium, and thereby, the individual QoS requirements of the coexisting networks are considered. Further, the SLS allows an optimized usage of the available spectrum in that an operation in radio spectrum, which was originally licensed for other communication systems, is facilitated, as the SLS implies a search for unused spectrum as well as a release if it is needed again. The ability of the SLS to support QoS in the presence of other, competing, cognitive radios and the prevention of harmful interference to licensed radio systems are evaluated in the below.

This section is structured as follows. The principle of SLS as Medium Access Control (MAC)-based approach to cognitive radios is discussed as two general examples in Section 6.3.2; the SLS is applied (i) for the distributed coordination of reservations, and (ii) for the coordination of direct access to shared spectrum by multiple cognitive radios. The rationale and the basic algorithm of SLS applied in the time domain are discussed in Section 6.3.3, using the example of a single frequency channel. In general, SLS is not limited to one channel and can instead be used on multiple channels. An illustration through initial simulations and a convergence analysis of the SLS algorithm is given in Section 6.3.4. A time-frame-based interaction model for the evaluation of spectrum sharing scenarios is introduced in Section 6.3.5 QoS support in IEEE 802.11e and its limitations in coexistence scenarios are outlined in Section 6.3.6. The application of SLS for coordinating reservations when reusing TV bands while considering an incumbent radio system and in modifying the IEEE 802.11e Hybrid Coordinator Controlled Access (HCCA) is described in Section 6.3.7. Section 6.3.8 describes the modification of the Enhanced Distributed Controlled Access (EDCA) of IEEE 802.11e through the SLS for enabling coordination of opportunistic spectrum access without reservation. The interaction of radios using SLS, their interference with a primary radio system, and the time to reach a mutually agreed distribution of allocations is evaluated. The SLS's ability to enable QoS support in these two scenarios is evaluated in Section 6.3.9. The limitations of applying SLS for distributed QoS support are analyzed in Section 6.3.9.4.

6.3.1 Related Work

This section on SLS is based on several publications, while the rationale and algorithm of SLS is introduced by Berlemann and Walke (2005). SLS with reservation is examined by Berlemann et al., (2005b) using the example of WiMedia WPANs autonomously

*Reproduced with permission from: L. Berlemann, G. R. Hiertz, B. Walke, and S. Mangold, Cooperation in Radio Resource Sharing Games of Adaptive Strategies. In: IEEE 60th Vehicular Technology Conference, VTC2004-Fall, Los Angeles CA, USA, 26–29 September 2004, 3004–3009. © 2004 IEEE

coordinating their resource reservations. The SLS without reservations and its application in EDCA spectrum sharing scenarios is introduced by Berlemann *et al.* (2005d). In addition, this section introduces the application of SLS for the reuse of TV frequency bands (Berlemann *et al.*, 2006c). A machine-understandable specification for the SLS as a policy in DARPA XG Policy Language (DARPA, 2004a), in the context of policy adaptive cognitive radios is given by Berlemann and Mangold (2005) and Berlemann *et al.* (2005e) and in Section 6.3.10.

The idea of SLS is derived from the principle of waterfilling from the field of multi-user information theory and communications engineering. In a multiple transmitter and receiver environment, waterfilling is used to solve a mutual information maximization problem based on the singular-value decomposition of a channel matrix (Kasturia *et al.*, 1990). Through the application of a multicarrier modulation, the transmission power can be adapted to the transfer function of the radio channel (Cover and Thomas, 1991; Gallager, 1968). This view is extended by iterative waterfilling in the context of multiple access channels as analyzed in detail by Popescu (2002), Popescu and Rose (2004) Yu, (2002) and Yu *et al.* (2004). In the context of cognitive radios, iterative waterfilling is also identified by Haykin (2005) as an alternative to a game theoretic interaction in a distributed transmit power-control problem. This section considers the transfer of waterfilling from its application in information theory to the SLS as part of the medium access of spectrum sharing cognitive radios in the time and frequency domain.

6.3.2 Enabling Cognitive Radios

SLS realizes the secondary usage of spectrum. Vertical spectrum sharing is enabled by avoiding harmful interference to primary radio systems. Additionally, the usage of shared spectrum is coordinated in a decentralized way, by taking individual QoS requirements into account. The SLS aims at improved efficiency of spectrum usage and at support of QoS in distributed environments. The core idea of SLS is to allocate spectrum with deterministic and predictable patterns, for example, with medium access intervals that are periodically distributed in time. The predictability of device allocations facilitates the mutual coordination of spectrum usage between different cognitive radios, even in scenarios where dissimilar radio technologies cannot communicate with each other, but can detect emitted radio signal energy from each other.

Figure 6.6 illustrates a potential outcome from SLS in a fictitious Time Division Multiple Access (TDMA)/Frequency Division Multiple Access (FDMA) system. The spectrum is divided into different time slots along the x-axis, different frequencies are on the y-axis, and the relative fraction of an allocation on the total length of a time slot on the z-axis. The dark gray resource allocations result from SLS of one or multiple cognitive radios, while the fixed, light gray allocations result from cognitive radios QoS restrictions and cannot be shifted or belong to an incumbent communication system. These fixed allocations are not considered for SLS, as they are not flexible in their placement within time/frequency domain. The level of the smoothed allocations, i.e. used fraction of time from a slot length on the z-axis is, in the following, referred to as 'load level'. The starting times of allocations or empty slots, where the SLS is based, are called 'ground'. The ground corresponds to the plane formed by the x/y-axis. The SLS consequences, under the restrictions of the devices' QoS requirements, equally distribute free quantities of the transmission medium.

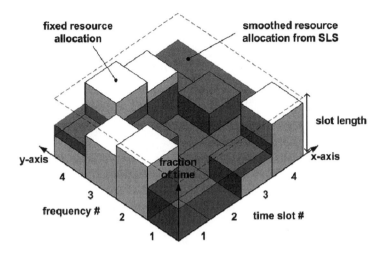

Figure 6.6 Spectrum load smoothing in the time and frequency domain of a fictitious TDMA/FDMA system. See also Figure 6.7 for an illustration of the SLS principle. Reproduced with permission from L. Berlemann, and B. Walke, Spectrum Load Smoothing for Optimized Spectrum Utilization – Rationale and Algorithm. In: IEEE Wireless Communication and Networking Conference, WCNC 2005, New Orleans LA, USA, 13–17 March 2005. © 2005 IEEE

6.3.2.1 Frame-based Interaction

Spectrum-sharing cognitive radios require a common temporal basis in order to facilitate an interaction. The point of time at which a cognitive radio acts in adapting transmission parameters has to be observable by all other spectrum sharing cognitive radios in order to enable a reaction to this action. All approaches discussed in this section have therefore a commonality, and the cognitive radios are synchronized in the sense that they interact on a common temporal basis. Here, the time between two consecutive beacons broadcast by the same (cognitive) radio establishes a time frame that provides an observable basis for interaction. Thus, the beginning and ending of an action can be clearly identified and a cognitive radio can react on it. In the application scenarios of this section, the time frame is introduced by the beacon interval, i.e. super frame, of IEEE 802.11e. The approaches presented in this section further assume, but do not require, simultaneous decision taking at the beginning/end of each frame.

6.3.2.2 Predictable Allocations as Contribution to Cooperation

Predictable allocations of a medium by a device enable an aimed interaction with other devices. They can therefore be regarded as a contribution to cooperation (Berlemann, 2002; Mangold, 2003), especially in the absence of a central coordinating instance like an AP or BS. Periodic resource allocations, i.e. deterministic and cyclic allocation patterns, of a device can be observed and predicted by all other spectrum sharing devices. These other devices may then adapt their own resource allocations with the aim of partially or completely preventing mutual interference on the shared medium. This can be considered as a further contribution to cooperation. The periodicity increases the possibility for other devices to conclude from the observed, delayed or after collisions, repeated resource allocations, on the originally demanded allocations of a device. These allocations correspond to the individual traffic demands and QoS requirements. The periodicity can be used as basis for introducing priorities as suggested for example by Walke (1978), there referred to as Rate Monotonic Priority Assignment (RMPA).

A further reduction of the period length, resulting from an increased number of equally distributed resource allocations per frame whilst the relative proportion of the resource allocations by devices per frame remains constant, may also be considered as a contribution to cooperation (Berlemann, 2002; Mangold, 2003). The aforementioned cooperation characteristics imply:

- interference reduction and avoidance;
- an increased chance for other devices to reduce the delays experienced for their data packets;
- a reduced blocking probability and access time for new devices initially accessing the medium.

Periodic resource allocations may preferably be performed during unused intervals of the frame to reduce the devices' mutual interference. Corresponding to the above-introduced aspects of cooperation, a cooperating device may improve its capability to support QoS if all other devices are cooperating as well. The definition of cooperation in the context of the game theory based approach in Section 5.5.5.1 takes these fundamentals into account.

6.3.2.3 Distributed Coordination of Reservations

Terrestrial TV broadcasts are currently being switched from analogue to digital, see Section 3.3.1. It is envisaged that an unlicensed reuse of the entire TV broadcast band will be allowed for cognitive radios that scan all TV channels throughout the band, and operate only upon identification of spectrum opportunities. In such a scenario, cognitive radios share a common resource (of secondarily used TV spectrum) and require mutual coordination to support QoS. The primary radio system (TV broadcast channels) has a quasi-fixed spectrum allocation that (i) can be identified by all cognitive radios, and (ii) founds the basis for their interaction.

One widely used approach to spectrum sharing is the usage of a Common Spectrum Coordination Channel (CSCC), which is outlined in Section 3.3.2. The basic idea of CSCC is to standardize a simple common protocol for periodically signaling radio and service parameters (Raychaudhuri and Jing, 2003). The DARPA XG Program (DARPA, 2004b) suggests a dedicated control channel located in licensed spectrum for coordination of spectrum sharing. We take up these ideas in applying SLS for a distributed coordination of reservations transmitted by cognitive radios on this dedicated control channel when reusing TV bands as outlined in Section 6.3.7.

6.3.2.4 Opportunistic Spectrum Usage

The SLS enables opportunistic spectrum usage through (i) identification of spectrum opportunities, (ii) using them in a coordinated way, and (iii) releasing the spectrum again if it is required by primary radio systems. Spectrum usage by opportunistic operation in licensed and unlicensed spectrum is illustrated in Figure 6.7. In this figure, characteristic spectrum usage patterns in the 5-GHz unlicensed band for three IEEE 802.11a frequency channels are shown, together with two adjacent channels in the licensed spectrum. Cognitive radios differ between three kinds of spectrum opportunities: (i) spectrum that is most of the time unused as it is reserved for radio systems that do not operate frequently, for example emergency or military services, (ii) deterministically used licensed spectrum, and (iii) predictably used unlicensed spectrum.

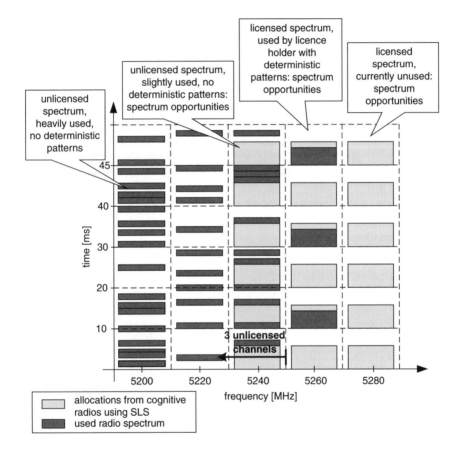

Figure 6.7 Spectrum usage example at 5 GHz. Three 802.11a channels and frequencies above are depicted. The dark gray fields indicate used spectrum. Cognitive radios use spectrum load smoothing to identify and allocate spectrum opportunities (light gray). Reproduced with permission from L. Berlemann, G. R. Hiertz, B. Walke, and S. Mangold, Cooperation in Radio Resource Sharing Games of Adaptive Strategies. In: IEEE 60th Vehicular Technology Conference, VTC2004-Fall, Los Angeles CA, USA, 26–29 September 2004, 3004–3009. © 2004 IEEE

The detection of such spectrum opportunities can be facilitated with the spectrum usage measurements of IEEE 802.11k (IEEE, 2005c; Mangold *et al.*, 2004a; Mangold and Berlemann, 2005). The example in Figure 6.7 illustrates the allocations (light gray) of spectrum sharing cognitive radios that apply the principle of SLS when allocating spectrum opportunities. IEEE 802.11a radios demand access to the channels at 5220 MHz and 5240 MHz with random patterns (dark gray). Cognitive radios distribute their allocations between the 802.11a allocations. As the 802.11a allocations cannot be reliably predicted by the SLS using cognitive radios, the cognitive radio's allocations might delay them 802.11a medium access attempts in according to the Carrier Sensing Multiple Access with Collision Avoidance (CSMA/CA) of 802.11 though ongoing cognitive radio transmissions. At 5260 MHz, a license-holding primary radio system uses spectrum with deterministic patterns, respected and not interfered with by the cognitive radios. In this case, the SLS-using cognitive radios are able to distribute their allocations around the correctly anticipated 802.11a transmissions. At 5280 MHz, a sporadically used licensed spectrum with few spectrum accesses, the cognitive radios coordinate each other in applying the SLS.

6.3.3 Spectrum Load Smoothing in the Time Domain

The rationale of SLS is described in the following based on the work of Berlemann and Walke (2005). For a better understanding, the dimensions of the medium under competition are kept to a single frequency in the remainder of this section without restricting the general applicability of the SLS to multiple frequencies. In our initial step, a simplistic radio channel is assumed and the potential hidden station problem is ignored.

6.3.3.1 The Algorithm

Figure 6.8 illustrates the principle of SLS using the example of a single frequency in the time domain. A periodic, frame-based, MAC protocol provides the basis for coordination and interaction. In the application scenarios of this section, this MAC frame is referred to as the IEEE 802.11e beacon period. Alternatively, we use the (not 802.11 standard compliant) notion of a superframe. Once per frame, a cognitive radio determines the intended spectrum access with the help of the SLS. Here, the frame consists of four slots of equal duration (lengths) whereby a slot is a time interval during which the 'multiple access' occurs. The slot length is respected by all cognitive radios. All cognitive radios need to know *a priori* the slot structure or learn it from observation. In a distributed environment, the slot length can be identified with the help of the autocorrelation function of the observed allocation patterns (Mangold et al., 2002; Mangold, 2003). Coexisting legacy communication systems or protocol specific limitations may, however, lead to violations of the slotted structure. The SLS deals with such violations by regarding an ongoing allocation from the last slot as the first allocation of the current slot, and following thereafter the intended access order of smoothed allocations.

The SLS is an iterative algorithm, as it redistributes the allocations of a cognitive radio with the aim of getting an equalized – smoothed – overall utilization of the four slots (and thereby the complete frame), which is referred to as load level. The initial two steps of the iterative determination of the smoothed load level are shown in Figure 6.8 (and also 6.26).

The iterative distribution of the cognitive radios' allocations over the available slots considers the observed added allocations of all other cognitive radios from the past as having a common origin. In Figure 6.8 only one cognitive radio, namely device 2, is present as interferer to device 1. The initial load level of device 1 is increased stepwise, beginning with

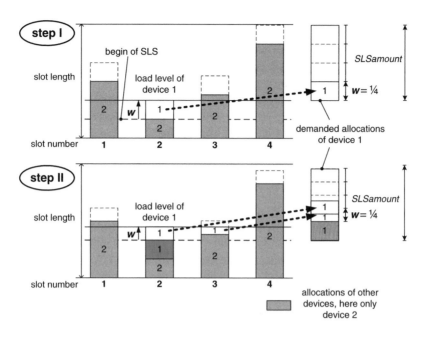

Figure 6.8 The principle of spectrum load smoothing in the time domain over a frame divided into four slots of the same slot length. The initial two iterative steps of the SLS are depicted. See also Figure 6.26 for an illustration of the SLS algorithm. Reproduced with permission from L. Berlemann, and B. Walke, Spectrum Load Smoothing for Optimized Spectrum Utilization – Rationale and Algorithm. In: IEEE Wireless Communication and Networking Conference, WCNC 2005, New Orleans LA, USA, 13–17 March 2005. © 2005 IEEE

the lowest allocation of device 2, here located in slot 2. The step size w of increasing the load level is given by:

$$w = \frac{amount\ of\ allocations\ to\ be\ distributed}{number\ of\ slots} \tag{6.6}$$

The difference between the load level and the allocations of device 2 is filled with allocations of device 1 (see Figure 6.8, step II, slot 3). These (spectrum load) smoothed allocations are subtracted from the amount still to be distributed, depicted in Figure 6.8 in the upper right corner of each step. Thus, from iteration to iteration, the load level of device 1 rises, and the step size w as well as the remaining amount of allocations decreases, when assuming a static amount of allocations from device 2. The accuracy of the algorithm defines a criterion for ending this iterative algorithm.

The cognitive radio's distributed allocations are placed in this example on the top – after – the allocations of the other cognitive radios. As all cognitive radios might (spectrum load) smooth their allocations simultaneously, rules for accessing a slot are necessary. Further, a broadcast of the intended allocations through reservations is preferable to prevent collisions and delays, as discussed and evaluated below. The SLS might imply a minimum and maximum size of an allocation after applying SLS. The reason for this can, for instance, be a reduction of the protocol overhead or restrictions to the transmission size depending on the coding and modulation scheme of the PHYsical layer (PHY).

6.3.3.2 Spectrum Load Smoothing With and Without Reservations

SLS can be categorized according to the usage of reservations: (i) SLS might be applied without reservation on the basis of observing past frames, and (ii) the accuracy of SLS can be improved through broadcasting reservations. The SLS without reservations is performed simultaneously by all cognitive radios at the beginning or end (which is the same) of a frame. In this case, the SLS is done stepwise from frame to frame. In redistributing a limited amount of allocations from the previous frame, a mutual interaction is enabled. Such a simultaneous and iterative SLS without reservations is considered below. In the case of reservations, i.e. a broadcasting of intended allocations for the actual frame, the SLS is done based on observed allocations of the past frame actualized through the reservations for the actual frame, if available.

The amount of allocations per frame considered for redistribution through SLS is called SLSamount. For SLS with reservations, all allocations can be shifted at once (SLSamount = 100%). To enable a fast, coordinated as well as stable, smoothed allocation scheme without reservations, the SLSamount is adapted, i.e. decreased, on the way to the smoothed allocation solution. Referring to control theory, the SLSamount can be regarded as attenuation factor. The flow chart in Figure 6.9 depicts, therefore, the SLS with and without reservations while the amount of redistributed allocations is flexible. The simulations, as introduced below, indicate that an initial value of SLSamount of 10 % is suitable for enabling stability and reaching a smoothed overall allocation distribution in a short duration of time.

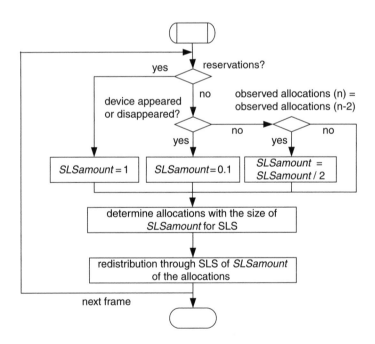

Figure 6.9 Flow chart of iterative SLS with adaptive amount of redistributed allocations. Reproduced with permission from L. Berlemann, and B. Walke, Spectrum Load Smoothing for Optimized Spectrum Utilization – Rationale and Algorithm. In: IEEE Wireless Communication and Networking Conference, WCNC 2005, New Orleans LA, USA, 13–17 March 2005. © 2005 IEEE

Before redistributing a specific amount of allocations through the SLS, the most destructive allocations on the way to a smoothed overall allocation scheme have to be identified. Destructive refers in this context to parts of allocations that are above the ideal, smoothed, common load level of all the slots. The identification of allocations that are to be cut is introduced in the next section. The SLSamount is halved, as outlined in Figure 6.9 and observable in Figure 6.13(b), if the overall allocations of the last but one frame equal the allocations of the present frame. In a yo-yo like manner, as depicted in Figures 6.12(d) and (e), the cognitive radios shift allocations at the same time to less utilized slots, overload these slots together, and shift in the consecutive frame their allocations back to the original slots. This effect is countered by decreasing the amount of redistributed allocations. In the case of a cognitive radio initiating or ending transmissions, the smoothed, mutually agreed, allocation solution is obsolete and has to be coordinated again. Therefore, the SLSamount is reset in this case back to 10 %.

For the sake of a better predictability, it is assumed that all cognitive radios cooperate as introduced above. They change their allocation scheme much less frequently than the frame frequency during ongoing transmission to ease mutual load coordination. This may be achieved through, for instance, an aimed buffering in the MAC layer.

6.3.3.3 Identification of the Most Destructive Allocations

Parts of the allocations from the previous frame, which are to be redistributed through SLS, are identified by reversing the SLS algorithm, as outlined for device 1 in Figure 6.10. A virtual

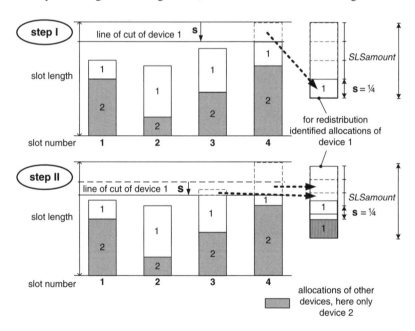

Figure 6.10 Determination of allocations that are to be shifted in the SLS. The SLS from Figure 6.8 is reversed and the first two steps are depicted. Reproduced with permission from L. Berlemann, and B. Walke, Spectrum Load Smoothing for Optimized Spectrum Utilization – Rationale and Algorithm. In: IEEE Wireless Communication and Networking Conference, WCNC 2005, New Orleans LA, USA, 13–17 March 2005. © 2005 IEEE

line of cut is iteratively moved down from the most utilized slot. The outstanding parts are cut and redistributed through the following application of SLS. The amount of cutting of a slot length, depicted in the upper right corner of each step, is given by the SLSamount. The line of cut with the aim of SLS is moved down with a step size s given by:

$$s = \frac{left\ amount\ of\ allocations\ to\ be\ cut\ for\ SLS}{number\ of\ slots} \tag{6.7}$$

The allocations identified for redistribution are summed up. In subtracting these allocations from the intended amount of allocation for redistribution, the remaining quantity defines the step size s of the next iteration corresponding to Equation (6.7). The accuracy of the algorithm defines again a criterion for ending this iterative algorithm.

6.3.4 Initial Simulations and Convergence Experiments

This section concretizes and initially evaluates (i) the SLS on the basis of reservations, and (ii) the SLS without reservations, using an adaptive amount of redistributed allocations from above. In the following, the SLS is now performed by three cognitive radios (device 2, 3 and 4), a frame structure of four time slots is assumed, and interactions over 75 frames are evaluated. These three cognitive radios operate on the same frequency channel and location together with an additional cognitive radio (device 1). This cognitive radio has a fixed allocation scheme, for instance, fixed allocations may result from an incumbent communication system not using SLS, as introduced in the opportunistic spectrum usage scenario of Section 6.3.8. A dedicated, protected, coordination phase where the reservations of the SLS-using cognitive radios are broadcast may also be the reason for fixed allocations as outlined in the WPAN scenario of Berlemann *et al.* (2005b).

6.3.4.1 SLS on the Basis of Reservations

The SLS based on reservations can be realized within the specific period of time used for coordination. This coordination phase can be realized with a dedicated coordination channel, or might be a reserved and protected part of the MAC frame. Within this coordination phase, the cognitive radios successively broadcast and coordinate their reservations in applying the SLS. The SLS thereby considers the already received reservations of the other cognitive radios, if available. Otherwise the observed and therefore less actual allocations of the last frame are taken into account. Such a dedicated coordination period embedded in a MAC frame is, for instance, part of the Mesh Networks Alliance proposal for 802.11s by Hiertz *et al.* (2005c).

Figure 6.11(a) depicts the observed normalized throughput of the three SLS using cognitive radios over time. Initially, device 2 demands a normalized share of capacity of 0.3, while device 3 demands 0.2. Specific events in the interaction route are marked with numbers and the corresponding allocation situations are depicted in Figures 6.11(b) to (d). The figures outline the demanded and observed allocations of the four time slots of the frame. For the SLS, it is assumed that the maximum load level of a time slot, i.e. considered maximum capacity, has a value of 0.8. The remaining capacity is essentially left unallocated to enable, for instance, access by additional SLS using cognitive radios or legacy devices. The maximum load level

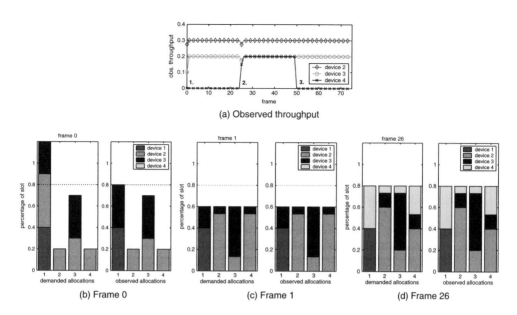

(a) Observed throughput

(b) Frame 0   (c) Frame 1   (d) Frame 26

Figure 6.11   Observed throughput (a) and allocations during the SLS (b–d). Device 1 has fixed allocations. The SLS is done by devices 2, 3 and 4. Device 4 initiates transmission in frame 25. All allocations
can be redistributed by the devices from frame to frame (SLSamount = 100%). Reproduced with permission from L. Berlemann, and B. Walke, Spectrum Load Smoothing for Optimized Spectrum Utilization –
Rationale and Algorithm. In: IEEE Wireless Communication and Networking Conference, WCNC 2005,
New Orleans LA, USA, 13–17 March 2005. © 2005 IEEE

is respected by all cognitive radios and they cut their allocations, i.e. abort their transmission,
if it is reached. The SLS is done over four time slots that cover the complete frame.

At the initial frame 0, marked by '1' in Figure 6.11(a) and depicted in Figure 6.11(b),
devices 1, 2 and 3 share the medium and their demanded allocations are uncoordinated: They
overload the first time slot leading to a shortened observed allocation for device 2 and no
allocation in this slot for device 3, implying less observed throughput as demanded. The
SLS already leads in frame 1 to a mutually coordinated demand for allocations, implying
a fulfilled demanded throughput for both cognitive radios as depicted in Figure 6.11(c). The
cognitive radios may redistribute all (SLSamount = 100%) of their allocations simultaneously
per frame, corresponding to the above-introduced SLS with reservations algorithm.

A fourth cognitive radio initiates transmission in frame 25, demanding 0.2 as share of
capacity and initiates its allocations at frame 25, see '2' in Figure 6.11(a), leading again
to an uncoordinated distribution of allocation. Some slots are overloaded, resulting again
in a reduced observed throughput. As all cognitive radios follow the SLS, devices 2 and
3 as well as 4 redistribute their allocations. The emerging outcome of SLS in frame 26 is
depicted in Figure 6.11(d). At frame 50, device 4 terminates its transmissions '3', resulting in
a redistribution of the allocations of the remaining cognitive radios as shown in Figure 6.11(c).

The emerging steady point of interaction can be regarded as a Nash Equilibrium from the
perspective of game theory. In focusing on the throughput, no cognitive radio can gain a higher

throughput in deviating from this solution. Although the cognitive radios still redistribute their allocations due to the SLS, the resulting allocation outcome is fixed and stable. The principle of the Nash Equilibrium is introduced in Section 5.5.8.1, and is also considered in the context of iterative waterfilling by Haykin (2005).

6.3.4.2 SLS Without Reservations

Without reservations the SLS has to be based on less accurate information. The observed allocations of past frames are considered for determining the expected allocations of other cognitive radios in the current frame. These observed allocations form a basis for the simultaneous SLS, done preferably at the beginning/end of the actual frame. Nevertheless, to enable coordination the redistribution process of allocations due to the SLS has to be slowed down for signaling purposes, as described above.

Figures 6.12 and 6.13(a) depict, analogously to Figure 6.11, the observed throughput and corresponding allocations during the interaction. Unlike the case of Figure 6.11, the amount of allocations is here adapted during the course of interaction following the SLS algorithm,

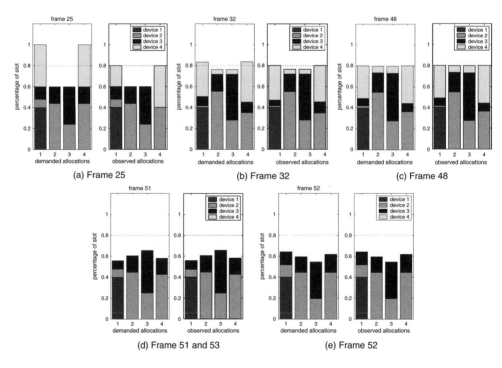

Figure 6.12 Allocations during the SLS. The amount of the allocations that are redistributed per frame is adapted as depicted in Figure 6.9. Reproduced with permission from L. Berlemann, and B. Walke, Spectrum Load Smoothing for Optimized Spectrum Utilization – Rationale and Algorithm. In: IEEE Wireless Communication and Networking Conference, WCNC 2005, New Orleans LA, USA, 13–17 March 2005. © 2005 IEEE

as introduced in the flow chart in Figure 6.9. The amount of allocations that are redistributed during one frame is decisive for the smoothness of the stable allocation scheme resulting from SLS. Figure 6.13(b) depicts the amount of shifted allocations per frame. All cognitive radios initiate their SLSamount with 0.1 and reset to this value if any cognitive radio appears (1), disappears (4) or rapidly changes its demanded allocations.

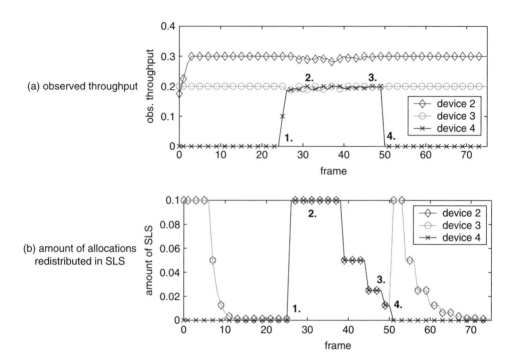

Figure 6.13 (a) The observed throughput of SLS, carried out by three devices, with the adaptive amount of redistributed allocations from frame to frame. (b) The amount of allocation, redistributed in the SLS, is adapted by the devices. Reproduced with permission from L. Berlemann, and B. Walke, Spectrum Load Smoothing for Optimized Spectrum Utilization – Rationale and Algorithm. In: IEEE Wireless Communication and Networking Conference, WCNC 2005, New Orleans LA, USA, 13–17 March 2005. © 2005 IEEE

The influence of the adaptive SLSamount is illustrated in Figure 6.12(a–c). The appearance of device 4 in frame 25, Figure 6.12(a) and '1', leads to an uncoordinated allocation distribution within the frame. The three cognitive radios shift 10 % (SLSamount = 10%) of their allocations stepwise, leading to the situation in Figure 6.12(b) and '2'. Thereafter, the SLSamount is halved until a predefined minimum, here 0.001, is reached leading to nearly ideal smoothed allocations in frame 48, Figure 6.12(c) and '3'. The above-mentioned yo-yo like shifting of allocations that motivate the adaptive SLSamount is outlined in Figure 6.12(d) and (e). The cognitive radios' distribution of allocations in frame 51 and 53 is equal and triggers a reduction in SLSamount, as depicted in Figure 6.13(b) according to the flow chart in Figure 6.9.

In summary, the introduction of the adaptive amount of redistributed allocations during the simultaneous, iterative, SLS moderates the inaccuracy of the SLS resulting from the missing reservation information. Nevertheless, it takes more time, compared with SLS with reservations, until a coordinated solution is reached as can be seen in a comparison of Figures 6.11 and 6.13(a).

6.3.5 Modeling Spectrum Load Smoothing in Spectrum Sharing Scenarios

In the following, a frame-based coordination model is defined to analyze and evaluate the application of the SLS, that is similar to the game model from Section 5.5.4. The medium access of IEEE 802.11e is extended and modified to realize the SLS and assess the resulting interaction when sharing spectrum.

6.3.5.1 Redistribution of Allocations by SLS

Figure 6.14 depicts the SLS in the time domain based on a slotted, periodic frame (the definitions are used below). Here, three decentralized operating cognitive radios coordinate their allocation of spectrum. Each cognitive radio performs SLS, i.e. distributes its allocations over a distance of smoothing introduced by the maximum tolerable service time (which is also defined below) of the cognitive radio's applications. The timing diagram of the resulting

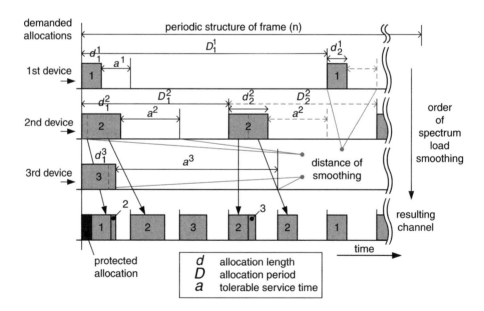

Figure 6.14 SLS in the time domain. Each cognitive radio (device) has an individual distance of smoothing. The periodic allocations are distributed so as to attain an overall smoothed spectrum usage. Reproduced with permission from L. Berlemann, G. R. Hiertz, B. Walke, and S. Mangold, Cooperation in Radio Resource Sharing Games of Adaptive Strategies. In: IEEE 60th Vehicular Technology Conference, VTC2004-Fall, Los Angeles CA, USA, 26–29 September 2004, 3004–3009. © 2004 IEEE

channel is also depicted. A cognitive radio decides about its allocation distribution at the beginning of the frame. This decision cannot be modified within the frame. The distance of smoothing is a multiple of the slot length, corresponding to the slotted structure of the frame that is introduced by device 1 as the first cognitive radio initiating a transmission. The order of SLS and thus the order of access to each slot are given through the temporal appearance of the cognitive radios. The protected allocation, shown in dark gray in Figure 6.14, may be from a primary radio system or represent a dedicated coordination period. Such a coordination period can be used for broadcasting reservations and is used for beacon broadcasts in the WiMedia WPAN scenario introduced by Berlemann *et al.* (2005b).

6.3.5.2 Definitions

The following definitions, illustrated in Figure 6.15, correspond to those of the game model introduced in Section 5.5.5.1. The model introduces a frame-based interaction consisting of three phases: (i) the decision about the intended allocations during the current frame corresponding to the SLS; (ii) the allocation of the shared medium under competition, and (iii) at the same time the observation of spectrum utilization as the basis for the SLS in the next frame. Figure 6.15 describes the QoS parameters of the frame-based interaction model. The SLS regards the required (upper) allocations as basis and results into demanded allocations per frame (center). The required allocations may be fragmented in order to fit adequately, corresponding to the SLS into the time slots. Due to the competitive access to the medium, the demanded allocations are interfered with, leading to delayed observed allocations (bottom).

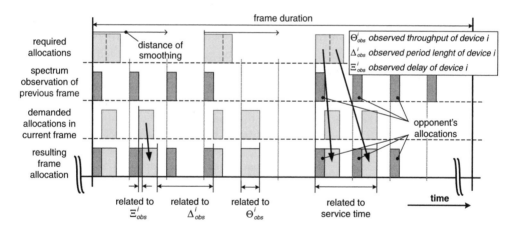

Figure 6.15 SLS in the time domain. The SLS regards required allocations as origin (top). QoS parameters result into demanded allocations per frame (center). Due to the competitive access to the medium the demanded allocations are interfered leading to delayed observed allocations (bottom)

Three abstract and (to the frame duration) normalized representations of QoS targets are defined in the following in the context of the coordination model, with the help of Figures 6.14 and 6.15: (i) The throughput $\Theta \in [0, 1]$, (ii) the period length $\Delta \in [0, 0.1]$ and (iii) the delay $\Xi \in [0, 1]$. The supported applications of the cognitive radios define the requirements for these

QoS targets. The definition interval of the period length reflects the arbitrary (but cooperation facilitating) assumption, that at least 10 allocation attempts per frame are intended.

The normalized throughput $\Theta^i(n)$ represents the share of capacity a cognitive radio i demands in frame n, and is defined as:

$$\Theta^i(n) = \frac{1}{FrameLength} \sum_{l=1}^{L^i(n)} d_l^i(n) \in [0,1] \qquad (6.8)$$

$L^i(n)$ is the number of allocations per frame n and $FrameLength$ the duration of the frame. The parameter $d_l^i(n)$ describes the duration of an allocation $l, l = 1 \ldots L$, of cognitive radio i in frame n. The normalized period length $\Delta^i(n)$ specifies the time between two consecutive allocations:

$$\Delta^i(n) = \frac{1}{FrameLength} max \left[D_l^i(n) \right]_{l=1..L^i(n)-1} \in [0,0.1] \qquad (6.9)$$

The period length is observable by all cognitive radios and plays an important role in distributed QoS support. The period length can be estimated by other cognitive radios and is regarded as a contribution to cooperation as described above in Section 6.3.2.2. In this way, the period length enables predictability and thus increases the success of mutual coordination (without reservations). The normalized observed delay $\Xi^i(n)$ is defined as the difference between demanded and observed allocation point of time and is part of the QoS evaluation below. The jitter can be directly derived from this observed delay but is not considered here.

The application of SLS as MAC layer-based approach for distributed coordination leads to an additional segmentation of allocations. Therefore, the service time, as time needed for completely transmitting all segments of a required allocation (e.g., an IP packet), is evaluated as a fourth QoS parameter. The services time refers to the total duration required for completely transmitting a higher layer data packet from one MAC layer entity to another, including the segmentation and reassembly functions. The service time thus reflects the overall transmission duration of a higher layer packet together with all delays and collisions that any fragment of that packet observed during its transmission over the shared medium.

Below in Section 6.3.9, the service time is used as defined in Figure 6.15 to enable a fair comparison with legacy scenarios in which SLS is not applied. The duration of an allocation, in 802.11e referred to as Transmission Opportunity (TXOP) duration, is reflected in the service time. The tolerable service time $d(n)$ is the maximum service time that the cognitive radio i tolerates in frame n and is introduced above as the distance of smoothing. Allocation attempts, which would lead to higher service times than the tolerable service time, are discarded.

6.3.6 QoS Support in IEEE 802.11e Coexistence Scenarios

In order to support QoS, IEEE 802.11e introduces a central instance referred to as Hybrid Coordinator (HC). The distributed, contention-based, channel access of the HC is referred to as EDCA. For a detailed description and evaluation of IEEE 802.11e see for instance Mangold (2003) and Mangold et al. (2003a). There it is shown that mutual coordination is desirable for QoS support on the basis of the EDCA in order to avoid collisions. The competitive access to each slot in the periodic frame is harmonized by the SLS by considering observed past frames. Collision avoidance is here intended to be reached by defining access order mechanisms to the wireless medium.

The following sections evaluate, in different spectrum sharing scenarios, the level of supported QoS. Spectrum sharing scenarios of completely overlapping networks that operate at the same frequency, time and location are considered. Further, side effects resulting from the hidden-station problem, link adaptation and power control are neglected and a simplistic radio channel is assumed. A decision making instance, as part of the Station Management Entity (SME), realizes the SLS approach in the protocol stack of IEEE 802.11e (Mangold, 2003). The frame-based coordination model and the basic IEEE 802.11e access mechanisms for a shared radio resource (here a common used frequency channel in the time domain) are evaluated with the help of the Matlab-based simulator YouShi, which is introduced in detail in Appendix B.

The behavior of a HC in the case of a collision is not precisely standardized; it is therefore assumed that the HC falls back to the EDCA in order to resolve the congestion. Figure 6.16 illustrates the QoS results, corresponding to the definitions above, of three coexisting HCs (HC0, HC1 and HC2) sharing the same frequency channel. The normalized observed throughput $\Theta^i(n), i \in 0\ldots2$ (upper part of figure), the observed period length $\Delta^i(n), i \in 0\ldots2$ (middle) and the observed maximum delay $\Xi^i(n), i \in 0\ldots2$ (bottom) of frame n are depicted. Figures 6.18 and 6.20 are structured in the same way and illustrate similar coexistence scenarios that are discussed later. The mutual interference of the HC's allocation attempts is evaluated over 15 IEEE 802.11e superframes. Each frame has a typical duration given by FrameLength = SFDUR = 100 ms. The QoS requirements of the throughput and period length are shown in gray. In the scenarios of this evaluation, the three 802.11e EDCA stations have the fixed requirement of allocating 20 % of the medium: $\Theta^i_{req} = 0.2, i \in 0\ldots2$. The requirements for the period lengths are assumed as follows: $\Delta^0_{req} = 0.1$, $\Delta^1_{req} = 0.1$ and

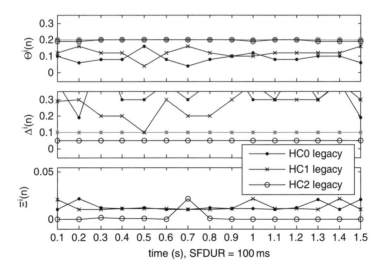

Figure 6.16 Legacy IEEE 802.11e HCCA coexistence scenario. Allocation attempts are uncoordinated and fail in colliding. A QoS guarantee is impossible

$\Delta_{req}^2 = 0.05$. In the legacy HCCA coexistence scenario of Figure 6.16, the allocation attempts of the HCs collide frequently, mutually delay each other and/or have to be discarded. Thus, the observed throughput is reduced and does not fulfill the requirement. The observed period length, i.e. the distance between allocation attempts, indicates that many allocations have been randomly delayed and discarded corresponding to the random backoff after collision of the EDCA. This leads to unpredictable allocations of the shared medium and thus illustrates the inability of legacy HCs to guarantee QoS without exclusive access to a shared medium.

6.3.7 SLS with Reservations – Approach to the Reuse of TV Bands

As outlined in Section 6.3.2.3, a dedicated coordination channel is one option for realizing secondary spectrum usage through signaling. The successful reception of a periodic signaling beacon indicates spectrum opportunities to the cognitive radios. This dedicated signaling beacon marks the coordination channel and could be transmitted from by a service provider in accordance with the operated-assisted cognitive radio approach from Section 3.3.3. In the case of the periodic beacon not being present, it is missed or else the beacon itself prohibits spectrum access, the cognitive radios defer immediately from spectrum access. In this way, an instantaneous release of spectrum is guaranteed when the primary radio system requires spectrum usage. The dedicated coordination channel is protected against interference by using a fraction of the unused licensed spectrum of the license holder designated for secondary usage.

As illustrated in Figure 6.17, modified IEEE 802.11e HCs can realize secondary spectrum usage in that HCs transmit their beacons sequentially on the coordination channel in the order of their local appearance. These beacons are extended standard beacons and contain

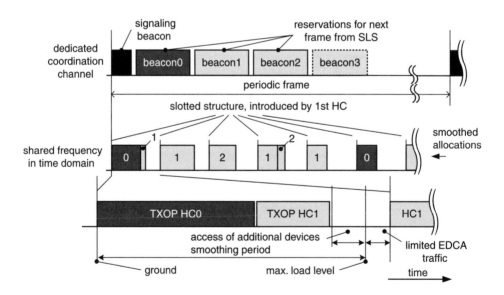

Figure 6.17 Reuse of TV-bands in an IEEE 802.11e HC spectrum-sharing scenario. HCs observe a dedicated coordination channel for signaling beacon allowing secondary usage. SLS is applied to coordinate spectrum access with reservations transmitted on the coordination channel

piggybacked reservations for spectrum access to frequencies available for secondary usage. These available frequencies are broadcast in the signaling beacon. The HCs apply the SLS to coordinate their reservations in a distributed way. Figure 6.17 illustrates as an example a single frequency shared by multiple HCs. The composition of the MAC frame and its timing diagram are depicted. The periodic beacon introduces a periodic frame structure for mutual coordination of the HCs; the points of time of the signaling beacons transmitted on the coordination channel are the basis for the frame-based interaction in the shared frequency. The reservations transmitted on the coordination channel refer to the subsequent frame (and not to the ongoing one). The HCs observe each frame and redistribute thereafter their demanded allocations for the next frame and adapt their broadcasted reservations accordingly.

Besides television broadcasts, TV broadcasting companies often operate additional communications systems in under-utilized TV bands assigned to them. Proprietary wireless communications systems are used to connect television cameras and microphones with outside broadcast vans. The nature of the SLS enables the prioritization and protection of such communications systems in case of spectrum sharing. The primary communication system, represented in Figure 6.17 by HC0, may introduce the slotted structure of the periodic frame corresponding to its QoS requirements and allocate spectrum accordingly. The SLS using HCs, here HC1 and HC2, coordinate their reservations by taking HC0 into account and distribute their allocations around HC0's allocations. The presence of the prioritized HC0 is not required for the SLS of the other HCs and they are able to coordinate reservations without any help from HC0.

A spectrum-sharing scenario of one primary radio system, here HC0, and two HCs (HC1 and HC2) using SLS with reservations for mutual coordination is depicted in Figure 6.18. The QoS requirements are the same as in the legacy HCCA coexistence scenario from above. The allocations of the primary radio system are to be prioritized with SLS using HC1 and HC2

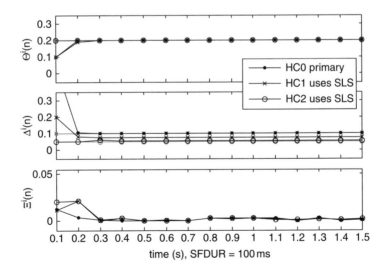

Figure 6.18 IEEE 802.11e HCCA spectrum sharing scenario of protected primary HC and two HCs using SLS with reservations. Although less prior EDCA is present, the SLS is successful. A stable coordinated allocation distribution is reached after three frames

to identify free time intervals and distribute their allocations around the transmissions from the license holding HC0. The EDCA traffic represents the EDCAs of all HCs and accesses the medium with less priority, i.e. longer waiting time before access, in case the medium is idle. The offered EDCA traffic load is 2 Mb/s and the TXOP duration of the EDCA's allocations is limited to TXOPlimit = 0.3 ms. Corresponding to the application example introduced above, the reservations are successfully transmitted in a dedicated coordination channel that the EDCA is not allowed to access. The presence of the EDCA has no influence on the success of the SLS and mutually coordinated distribution of allocations is reached after three frames. Nevertheless, the EDCA's allocations result in extended period lengths $\Delta^i(n)$ and delays $\Xi^i(n)$ for all HCs.

6.3.8 SLS Without Reservations – Opportunistic Spectrum Usage Scenario

The opportunistic spectrum usage in applying SLS without reservations is outlined in this section. The timing diagram of a periodic IEEE 802.11e superframe in Figure 6.19 illustrates the medium access of cognitive radios corresponding to a modified 802.11e EDCA in order to avoid mutual delays and collisions. The 802.11e superframe has again a slotted structure, here introduced by EDCA1, which is used for applying the SLS as introduced in Section 6.3.3. The primary radio system is represented by EDCA0, and its allocation attempts have to be successful and may not be interfered with by the cognitive radios (realized as modified EDCA1-4). These SLS using EDCAs follow a coordinated order of access to prevent collisions. The sequential order of access possibilities for each EDCA through periods intentionally left free, common to all slots of the frame, is given by the order of initial transmission of the EDCA within the considered local coverage area. As illustrated in Figure 6.19, the SLS can be done, without any announcement of reservation information, with the help of an individual access period for each cognitive radio in each slot. The period of opportunity to access the medium is

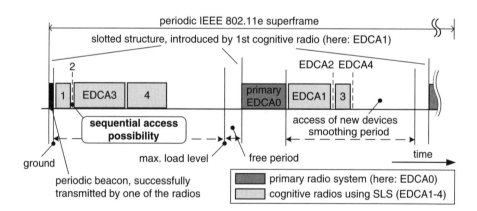

Figure 6.19 IEEE 802.11e EDCA coexistence scenario. SLS is used without reservations. The stations have an individual access period within a slot in the order of their initial transmission. Reproduced with permission from L. Berlemann, G. R. Hiertz, B. Walke, and S. Mangold, Cooperation in Radio Resource Sharing Games of Adaptive Strategies. In: IEEE 60th Vehicular Technology Conference, VTC2004-Fall, Los Angeles CA, USA, 26–29 September 2004, 3004–3009. © 2004 IEEE

left free in each slot for each cognitive radio independently from the real demanded allocation of a certain slot. In accordance with the SLS principle, no communication for coordination is required between the primary radio system and the SLS using cognitive radios with their modified EDCA. Each cognitive radio observes the allocations of the past frames and identifies future time periods in which the primary radio system is not accessing the medium. It is assumed that the allocation patterns of all spectrum-sharing radios do not fluctuate much from one frame to another and an additional buffering is done in order to enable deterministic allocation patterns.

In the coexistence scenario of IEEE 802.11e, for stations using EDCA in this section the SLS is performed simultaneously at the beginning/end of each frame. A spectrum-sharing scenario of one incumbent primary radio system, here EDCA0, and two EDCAs that apply SLS without reservations for mutual coordination is depicted in Figure 6.20. The QoS requirements are the same as in the legacy HCCA coexistence scenario from above. The allocations of the primary license-holding radio system are to be protected: The SLS using EDCA1 and EDCA2 identify free time intervals and distribute their allocations around the transmissions from the incumbent EDCA0. The slotting for SLS is introduced by the periodic allocations of EDCA0. The transmission interval is observable and can be identified by EDCA1 and EDCA2 with the autocorrelation function of the allocation patterns (Mangold *et al.*, 2002; Mangold, 2003). The frame is again divided for SLS into 40 slots and it is assumed that EDCA1 has a fixed distance of smoothing (maximum tolerable service time) of $a^1 = 7.5$ ms, while EDCA2 has a distance of smoothing of $a^2 = 5$ ms. It can be seen from Figure 6.20 that the first frame is required for EDCA1 and EDCA2 to observe the allocation pattern from EDCA0. After the second frame, interference to EDCA0 is avoided. An observation of EDCA0 before initially accessing spectrum would prevent this interference but it cannot be assumed in general, that the primary radio system is already transmitting, when cognitive radios would like to access spectrum opportunistically.

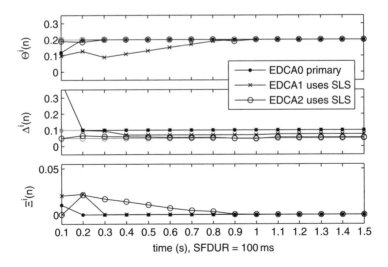

Figure 6.20 IEEE 802.11e EDCA spectrum-sharing scenario. A primary radio system, here EDCA0, is to be protected and two secondary EDCAs use SLS without reservations. A coordinated distribution of allocations is reached after 11 frames

In the HCCA scenarios above, all allocations can be redistributed per frame (SLSamount = 100%) due to the usage of reservations. Alternatively, this spectrum-sharing scenario of EDCA stations implies a more complicated coordination problem as outlined in Section 6.3.2.4, where, as no reservations are used, the SLS is based here on the less accurate observation of past frames. An adaptive amount of allocations considered for SLS is required to enable a convergence of the simultaneous redistribution of allocations, see Figure 6.9. In order to focus on the main effects, a constant SLSamount of 10 % is assumed in the following, which is adequate for reaching a stable allocation distribution in a short time (Berlemann and Walke, 2005). Due to the missing information about current frame allocations, 11 frames are required in this scenario in order to reach a coordinated (stable and smoothed) solution. Nevertheless, EDCA0 is not interfered with and all cognitive radios applying the modified EDCA observe their required throughput. Their allocations – fixed (EDCA0) or distributed by the SLS (EDCA1 and EDCA2) – do not collide or delay each other. The presence of the incumbent radio system increases the distance between two consecutive allocations and EDCA2 fails to meet its requirement $\Delta^2_{req} = 0.05$.

In comparing the results from SLS without reservation (Figure 6.20) with the results from SLS with reservations (Figure 6.18), the advantage of reservations is clearly seen. In two comparable scenarios, the usage of reservations leads after four frames to a coordinated solution in comparison with the 11 frames required without reservations. In both scenarios, the primary radio system is successfully protected from interference in applying the SLS principle.

6.3.9 Evaluation of QoS Capabilities

The three spectrum sharing scenarios from above are analyzed in this section in relation to the support of QoS. Figures 6.21, 6.22 and 6.23 depict therefore, the normalized observed throughputs and the complementary Cumulative Distribution Functions (CDF) of the observed service times of the three radios (either HCs, modified HCs or cognitive radios applying a modified EDCA) in the corresponding spectrum-sharing scenario. As above, the

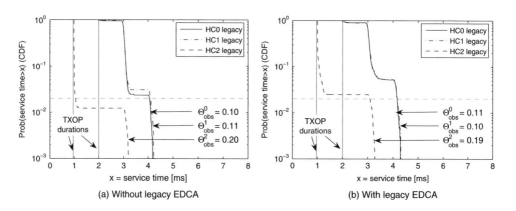

Figure 6.21 Service times and throughputs in the coexistence scenario of three 802.11e HCs. Due to missing coordination, the observed throughput is essentially reduced. The HC allocations considerably delay each other. In (b) there is, in addition, lower priority EDCA traffic present

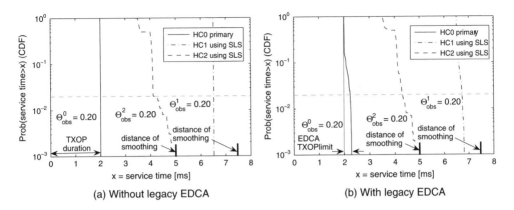

Figure 6.22 Service times and throughputs in the 802.11e HC spectrum-sharing scenario. The HCs use SLS reservations in transmitting them on a dedicated coordination channel. Contrary to part (a), lower priority EDCA traffic is present in (b)

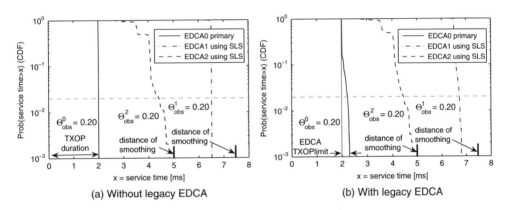

Figure 6.23 Service times and throughputs in the 802.11e EDCA spectrum-sharing scenario. The EDCA0 represents the primary radio system. EDCA1 and EDCA2 use SLS without reservations to coordinate opportunistic spectrum access. No legacy EDCA traffic is present in part (a)

radios require to access 20 % of the medium ($\Theta^i_{req} = 0.2, i \in 0 \dots 2$) and the required period lengths are assumed again to be $\Delta^0_{req} = 0.1$, $\Delta^1_{req} = 0.1$ and $\Delta^2_{req} = 0.05$. The QoS evaluation is performed over 200 frames (with SFDUR $= 100$ ms), while the initial frame required for mutual observation, as illustrated in the figures considering 15 frames from above, is not considered here. Each figure depicts the QoS results without legacy EDCA traffic (a) and in the presence of EDCA background traffic (b). The lower priority EDCA traffic accesses the medium in the case of its being idle for a certain period of time. The offered EDCA traffic load is 2 Mb/s and the TXOP duration of the EDCA's allocations is limited to TXOPlimit $= 0.3$ ms. Due to its lower priority access and the presence of multiple HCs allocating the medium, the EDCA is overloaded here. The dashed gray horizontal lines mark the 98-percentile of the service times' CDF.

6.3.9.1 Legacy HCCA Coexistence

Figure 6.21 illustrates demonstratively the coexistence problem of legacy HCs sharing the same frequency. Both, HC0 and HC1 fail considerably to meet their required throughput due to frequent collisions, mutual delays, and/or discarding of intolerably delayed allocations, as also indicated in Figure 6.16. The fixed TXOP durations of the HCs of 1 ms and 2 ms respectively are observable, as they are contained in the service times. The steps in the CDF reflect the mutual delays and the edges are reasoned in collisions and resulting backoff procedures of the EDCA for contention resolution. The coexistence problem leads to unpredictable service times, which are out of the control of the HCs. The presence of legacy EDCA traffic leads to longer service times, as seen at the 98-percentile line in comparing parts (a) and (b) of Figure 6.21.

6.3.9.2 SLS with Reservations

The observed QoS in the HC spectrum-sharing scenario with the distributed coordination of reservations is shown in Figure 6.22. All HCs fulfill their required observed throughputs. HC0 has a fixed allocation pattern, while HC1 and HC2 separate their allocations and use SLS for redistributing them. The service time distribution of HC0 indicates in (a) that HC1 and HC2 are not delaying HC0, and in (b) that the legacy EDCA traffic delays the HC0 allocations up to its TXOPlimit. The fixed TXOP duration of the prioritized HC0 of 2 ms is also clearly seen. The distance of smoothing (for HC1 7.5 ms and for HC2 5 ms) is an upper limit for the maximum observed service time (transmission duration). Thus the tolerable service time, i.e. distance of smoothing, is decisive for the observed total service time. A completely interference-free operation of HC0 can be guaranteed if the EDCA is prohibited, although the EDCA TXOPlimit is already under control of the HCs.

6.3.9.3 SLS Without Reservations

The opportunistic access to spectrum under protection of a primary radio system is evaluated in Figure 6.23. EDCA0, representing the medium access of a primary radio system has a fixed allocation pattern, while EDCA1 and EDCA2 are harmonized in applying the SLS without reservations with SLSamount = 10% for mutual coordination as outlined above. A comparison of (a) and (b) indicates that no legacy EDCA should be permitted in order to prevent any interference to the primary radio system. The EDCAs of all radios fulfill their required throughputs and the primary radio system is not interfered with in (a) where the observed service time of the primary radio system using EDCA0 is reduced to the transmission duration. The distances of smoothing (the same as above) are again an upper limit for the maximum observed service time. A distributed coordination and deterministic spectrum access, as required for the support of QoS, is successfully reached in applying SLS.

6.3.9.4 Limitations of Spectrum Load Smoothing

The application of SLS enables QoS support in distributed environments on the basis of observing past spectrum utilization and/or reservations, as shown above. The necessary coordination is reached as all cognitive radios have the same target in using the SLS, namely the smoothed utilization level of the time slots. Severe QoS requirements result in distinct restrictions related to access time and access duration. The SLS provides means to shift and separate allocations so that they interlock similarly to the teeth of a comb in taking the tolerable service time into account. Nevertheless, when multiple cognitive radios require spectrum access simultaneously without tolerating any delays, the SLS fails. The nature of SLS, which provides deterministic idle times of the wireless medium, has thereby already solved this problem. A targeted single shifting of the required allocations times resolves the congestion.

The success of the SLS depends on the cooperation of all radios sharing the same spectrum, i.e. all radios apply the SLS. The main advantage of the SLS – its simplicity – leads to weakness against selfish, non-cooperative cognitive radio. Selfish cognitive radio demands more of the wireless medium than required in order to protect own allocations against interference from other radios, as described in Section 5.5.7.2. Contrary to the application of game theory from Section 5.5, the SLS has no means of enforcing cooperation through punishment.

Figure 6.24 illustrates this weakness by depicting the initial 15 frames in an EDCA spectrum-sharing scenario. The QoS requirements are the same as in Figure 6.16 and subsequently. Here, the radio applying EDCA0 is selfish in demanding a higher throughput than required $\Theta^0_{dem} = 0.52 \gg \Theta^0_{req} = 0.2$. In accordance with the SLS principle, the SLS using cognitive radios applying EDCA1 and EDCA2 free the slots, overloaded by the radio of EDCA0 and try to allocate less utilized slots. The observed throughput indicates that the selfish radio of EDCA0 is able to block out the cognitive radios of EDCA1 and EDCA2 within three frames. Both cognitive radios try to redistribute their allocations but are limited in this by their distance of smoothing. Consequently, allocation attempts are discarded due to intolerable expected service times. Therefore, the cognitive radios of EDCA1 and EDCA2 fail to meet

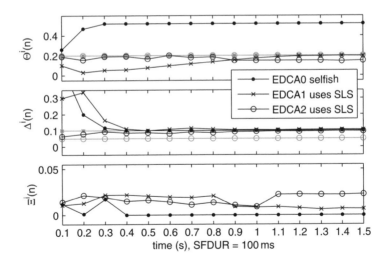

Figure 6.24 IEEE 802.11e EDCA spectrum-sharing scenario with a selfish radio present. EDCA0 is selfish in demanding a higher share of capacity than required. EDCA1 and EDCA2 are harmonized in applying SLS and corresponding allocations are blocked out of the medium by EDCA0. EDCA1 and EDCA2 considerably fail to meet their QoS requirements

their required throughputs ($\Theta^i(n)$; gray lines) and delays, as shown in Figure 6.24. The block-ing out is also indicated by the observed period length $\Delta^i(n)$. The cognitive radio applying EDCA2 in using the SLS observes greater distances between two consecutive allocations than required, as the allocations of the EDCA0 completely take over the medium for several slots. EDCA1 and EDCA2 consider EDCA0's allocations similarly to allocations from a primary radio system, any interference is avoided and overloaded slots (done on purpose by EDCA0) are freed independently from the own QoS requirements. The observed delays $\Xi^i(n)$, illus-trated in Figure 6.24, indicate the same, that is EDCA0's allocations are not delayed, contrary to those of EDCA1 and EDCA2.

The observed throughputs and service times over 200 frames in an EDCA spectrum-sharing scenario of one selfish EDCA0 and two SLS using EDCA1 and EDCA2 are depicted in Figure 6.25. The selfish EDCA0 observes its demanded throughput $\Theta^0_{dem} = 0.52$ and its allocations are not delayed. Due to the increase in demanded throughput EDCA0's TXOP duration is now 5.2 ms. On the other hand, for the SLS using cognitive radio, their allocations are considerably delayed, in the case of EDCA1 up to 6 ms and for EDCA2 up to 3.1 ms. Furthermore, the EDCA1 maximal service time is above its distance of smoothing of 7.5 ms. The EDCA0 allocations are again the reason for this. The allocations of EDCA1 are delayed beyond the designated SLS slot and continue into the next slot.

In summary, SLS-using cognitive radios require protection against selfish cognitive radios. This protection may be introduced through regulation of certain MAC parameters, such as a limit for allocation durations or a necessarily required distance between two consecutive allocations. The SLS from above fails as the TXOP duration of EDCA0 results into service times above the tolerable distance of smoothing for the other EDCAs. Alternatively, the SLS is an option for spectrum etiquette. It has been shown that when all cognitive radios sharing spectrum apply it, SLS enables QoS support in distributed environments.

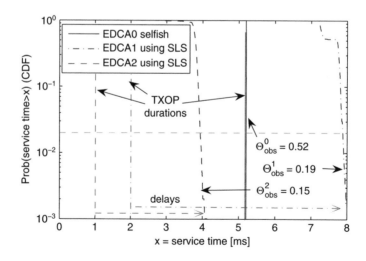

Figure 6.25 Service times and observed throughputs in the IEEE 802.11e EDCA spectrum-sharing scenario with a selfish radio present. EDCA0 is selfish in demanding a higher share of capacity than required. EDCA1 and EDCA2 are harmonized in applying SLS and corresponding allocations are blocked out of the medium by EDCA0. EDCA1 and EDCA2 considerably fail to meet their QoS requirements

6.3.10 Spectrum Load Smoothing as Policy*

The application of 'waterfilling,' a known principle in information theory, to the spectrum access of cognitive radio is referred to in Section 6.2.6 as SLS. In a few words, SLS realizes the secondary usage of spectrum. Vertical spectrum sharing is enabled by avoiding harmful interference to primary radio systems. With SLS, spectrum sharing cognitive radios aim simultaneously at an equally smoothed overall utilization of the spectrum. Figure 6.26 illustrates

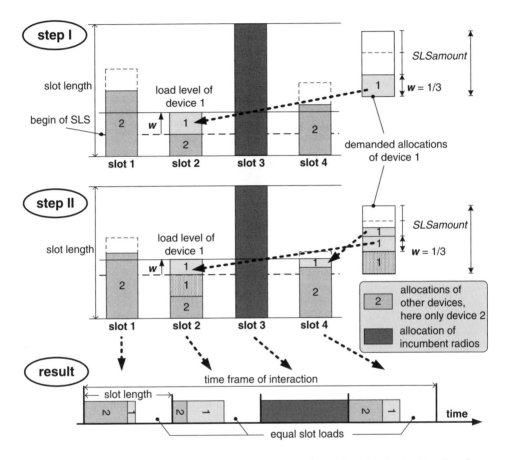

Figure 6.26 Illustration of the iterative spectrum load-smoothing algorithm in the time domain at a single frequency. The two initial steps are depicted together with the final result of the iterative algorithm. Reproduced with permission from L. Berlemann, S. Mangold, and B. H. Walke, 'Policy-based Reasoning for Spectrum Sharing in Cognitive Radio Networks,' in Proc. of 1st IEEE International Symposium on New Frontiers in Dynamic Spectrum Access Networks, DySPAN2005, Baltimore MD, USA, 8–11 November 2005. © 2005 IEEE

---

* Reproduced with permission from: L. Berlemann, S. Mangold, and B. H. Walke, "Policy-based Reasoning for Spectrum Sharing in Cognitive Radio Networks," in Proc. of 1st IEEE International Symposium on New Frontiers in Dynamic Spectrum Access Networks, DySPAN2005, Baltimore MD, USA, 8–11 November 2005.

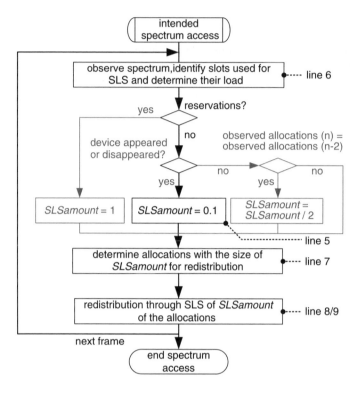

Figure 6.27 Flow chart of the spectrum load-smoothing algorithm. The black parts are described in Policy 6.3 and the respective lines are referred, while the gray parts are not considered here. Reproduced with permission from L. Berlemann, S. Mangold, and B. H. Walke, 'Policy-based Reasoning for Spectrum Sharing in Cognitive Radio Networks,' in Proc. of 1st IEEE International Symposium on New Frontiers in Dynamic Spectrum Access Networks, DySPAN2005, Baltimore MD, USA, 8–11 November 2005. © 2005 IEEE

the iterative determination of this smoothed load level. A flow chart of the steps to be performed when applying SLS algorithm is illustrated in Figure 6.27. This flow chart is the basis for specifying the SLS in Policy 6.3.

6.3.10.1 Aspects Relevant to Description as Policy

A periodic, frame-based, MAC protocol is the basis for coordination and interaction in applying the SLS. A frame is composed of time slots, i.e. detected spectrum opportunities, as depicted in Figure 6.26 at the example of one frequency that is shared in the time domain. The two initial steps of the SLS in determining the smoothed load level are depicted, together with the final result of this iterative algorithm. The time is divided here into four slots, building a time frame for interaction. The SLS is done over three slots (slots 1, 2 and 4), while slot 3 is not used in order to protect the incumbent radio transmission that is transmitting there. In a distributed environment, this slot length can, for example, be identified with the help of the autocorrelation function of the observed allocations, in case these are deterministic.

Policy 6.3 Spectrum load smoothing expressed in shorthand notation of the DARPA XG policy language. Reproduced with permission from L. Berlemann, S. Mangold, and B. H. Walke, 'Policy-based Reasoning for Spectrum Sharing in Cognitive Radio Networks,' in Proc. of 1st IEEE International Symposium on New Frontiers in Dynamic Spectrum Access Networks, DySPAN2005, Baltimore MD, USA, 8–11 November 2005. © 2005 IEEE

1   (PolicyRule (id SLS) (selDesc S1)(deny FALSE)
    (oppDesc TimeSlotsForSLS)(useDesc SLS2Slots))

2   (DeviceCap (id SLSProfile)(hasPolicyDefinedParams
    Observation.ownSpectrumUsage
    Observation.othersSpectrumUsage
    Observation.DevicesBefore
    Allocation.TransmitStart Allocation.AccessIntervals
    Allocation.AccessOrder)
    (hasPolicyDefinedBehaviors
    ObserveSlotStructure ObserveSlots
    SLScutAllocations SLSCalcLoadLevel))

3   (FrameDesc (id Observation.SlotsUsedForSLS)
    (Slot1 Slot2 Slot4)) /*from line 4*/
    (TimeDuration (id FrameDuration)
    (magnitude 50)(unit msec)) /*from line 4*/
    (Num (id Observation.DevicesBefore)
    (boundBy Device) (unit NONE))

4   (Process (id ObserveSlotStructure)
    (input Observation.ownSpectrumUsage
    Observation.othersSpectrumUsage)
    (output Observation.SlotsUsedForSLS FrameDuration))

5   (TimeDuration (id Observation.SlotLoad)(unit msec))
    (TimeDuration (id Allocation.SlotLoad)(unit msec))
    (SLSamount (id Allocation.SLSamount)
    (magnitude 10) (unit Percent))
    (TimeDuration (id Allocation.SLSLoadLevel)(unit usec))

6   (OppDesc (id TimeSlotsForSLS) (xgx '(and
    (invoke (before Allocation.TransmitStart)
    ObserveSlotStructure Observation.ownSpectrumUsage
    Observation.othersSpectrumUsage
    SlotDesc Observation.SlotsUsedForSLS
    TimeDuration FrameDuration)
    (invoke (within FrameDuration) ObserveSlots
    SlotsUsedForSLS TimeDuration Observation.SlotLoad)'))

7   (UseDesc (id SLS2Slots) (xgx '(and (invoke
    SLScutAllocations Allocation.SLSamount
    Observation.SlotLoad TimeDuration Allocation.SlotLoad)

(continued overleaf)

Policy 6.3   (*continued*)

---

8   (invoke SLScalcLoadLevel Allocation.SLSamount
    Allocation.SlotLoad TimeDuration
    Allocation.SLSLoadLevel)

9   (: = Allocation.AccessIntervals
    (− Allocation.SLSLoadLevel Allocation.SlotLoad)
    (: = Allocation.AccessOrder
    (+ Observation.DevicesBefore 1))′))

The smoothed level of utilization is determined in accordance with Figure 6.26. In Policy 6.3, it is referred to as SLSLoadLevel and calculated using the process SLSCalcLoadLevel. The access order of the cognitive radios to each slot is given by the order of a device's initial transmission within the considered coverage area. It is contained in Observation.DevicesBefore (line 3). To enable a convergence of interaction, the SLS is performed step-by-step from frame to frame. A limited amount of allocations is redistributed from one frame to another. The amount of allocations per frame considered for redistribution through SLS is called SLSamount. It has here a fixed value of 10 % (Berlemann and Walke, 2005) (line 5) corresponding to the flow chart in Figure 6.27.

A spectrum opportunity corresponding to the SLS is identified in two steps (line 6): firstly the slotted frame structure is defined through the process ObserveSlotStructure. Secondly, the usage, i.e. load, of each slot is observed by the process ObserveSlots from frame to frame. The decision as to how long and in which order to access slots within a frame is done in SLS2Slots (lines 7–9). Note that this decision about allocations is done simultaneously by all cognitive radios at the start of a frame.

6.3.10.2 SLS in the Policy Language

Policy 6.3 expresses the SLS in the shorthand notation of the DARPA XG policy language. The lines can be described as follows:

- Line 1 – The policy SLS specifies the usage of spectrum opportunities described in TimeSlotsForSLS corresponding to usage description SLS2Slots and selector description S1 defined in Policy 5.1.
- Line 2 – Parameters (hasPolicyDefinedParams) and processes (hasPolicyDefinedBehaviors) that a cognitive radio has to support in order to use the SLS policy are defined. The parameters are differentiated into (i) parameters used for observation to detect and specify spectrum opportunities provided by the cognitive radio, which have 'Observation.' as prefix; (ii) parameters used for specifying the allocation of the time slots resulting from the SLS policy with 'Allocation.' as prefix. These specify a dedicated spectrum usage for one frame. The processes ObserveSlots and ObserveSlotStructure are used in the opportunity description (line 4). SLScutAllocations and SLSCalcLoadLevel are used in the usage description SLS2Slots (lines 7–9).
- Line 3 – These definitions specify the frame structure and the access order of the cognitive radio. Observation.SlotsUsedForSLS is a list of time intervals. The slots are envisaged

as being used by the SLS, i.e. regarded as spectrum opportunities. Here, corresponding to Figure 6.26, Slot1, Slot2 and Slot4 are identified for SLS and have the type TimeInterval, which has two properties: starttime specifies the start of a time range and endtime the end. This multifield of time intervals results together with FrameDuration from the process ObserveSlotStructure (line 4). Observation.DevicesBefore is provided by the cognitive radio and enables determination of the order of access to spectrum in each slot.

- Line 4 – The process ObserveSlotStructure is defined for illustration purposes. Each cognitive radio that operates under the SLS policy has to implement this process and provide the grounding of the variables for it.
- Line 5 – A set of parameters is declared. The two '.SlotLoad' multifield variables consist of entries for each slot. Allocation.SLSamount is set to 10 %.
- Line 6 – An opportunity description named TimeSlotsForSLS. The slots used for SLS are observed during a frame with the ObserveSlots process. The list of time slots is the input parameter and the observed load for each slot is the output parameter of this process
- Line 7 – Usage description of the SLS. The allocations to be distributed in this step (the amount is defined by Allocation.SLSamount) are determined with SLScutAllocations.
- Line 8 – Thereafter, the load level Allocation.SLSLoadLevel is calculated using the process SLScalcLoadLevel.
- Line 9 – The Allocation.AccessIntervals for each slot are determined together with the order for accessing each slot Allocation.AccessOrder. The cognitive radio accesses spectrum in giving all the radios the higher priority of access that was already operating at the location when initiating own transmissions.

6.3.11 Learning from the Spectrum Load Smoothing Approach

SLS can be applied to other radio technologies than IEEE 802.11e. IEEE 802.16 for instance might be considered for an application of the SLS with reservations, when the Broadcast Channel (BCH) is considered as a reservation channel, coordinated in a distributed way with the help of the SLS, and the subscriber stations are able to inform the base station in the preceding frame on their intended allocations.

The SLS is a candidate approach in realizing QoS support in different licensing approaches to secondary spectrum usage and spectrum sharing. As an example, by applying SLS a restricted reuse of TV bands based on reservations can be coordinated. Further, the opportunistic access to under-utilized spectrum under the protection of a primary radio system is enabled in applying the SLS. The applicability of SLS is independent of the number of radio networks for completely/partially overlapping wireless networks.

The success of the SLS depends essentially on the cooperation by all radios, as no means to enforce cooperation such as punishment are considered. Thus the SLS requires regulatory protection or it is an option for a spectrum etiquette imposed by the license holder or regulation authority.

The specification of SLS in the policy language indicates the expressiveness of the policy framework. The SLS can be completely specified in the policy language and could now run on a cognitive radio capable of processing an XML-based spectrum policy. An experimental comparison with the game strategies as policies in the same language from Section 5.5.14 might be possible.

7

Our Vision – True Cognitive Radio

With this chapter, we round off our discussion of enablers for vertical and horizontal spectrum sharing from the previous chapters. We close the collection of arguments about the predominant research topics in the area of cognitive radio by looking forward into the future of wireless communication technology. So far we have been laying out some of the most relevant building blocks needed to realize a more flexible regulation with dynamic spectrum access. We have discussed regulation, standards, algorithms, and the main institutions currently working in this exciting, interdisciplinary, field of research. In the following, we will derive arguments from these trends and highlight our vision of what we will refer to as true cognitive radio (Berlemann and Mangold, 2005; Berlemann *et al.*, 2005b).

The well-known cognition cycle first published by Mitola (2000) will be discussed in the next section. In Section 7.2, we will continue with a brief review and critical statements about software defined radio, a concept that was developed in the hope of an improved flexibility in spectrum access.

We will illustrate in Section 7.3 how a policy framework that includes an Extendable Markup Language (XML)-based policy language can be used for cognitive radio and for specifying standards. As in the previous chapters, the discussion will be based on an early version of the Defense Advanced Research Projects Agency (DARPA) NeXt Generation Communication (XG) policy language (DARPA, 2003, 2004a, b).

Next, spectrum etiquette and operator-assistance will be promoted in the next two sections, 7.4 and 7.5. Some general thoughts on business opportunities in the field of cognitive radios will be discussed in the last section of this chapter, Section 7.6.

7.1 Mitola's Cognition Circle and Related Cognitive Radio Definitions

We support the thesis that true cognitive radio will make use of Mitola's early innovation, coined as 'cognition circle' (Mitola, 2000).

Cognitive radio will lead to a revolution in wireless communication, with significant implications for technology and regulation of spectrum usage in order to overcome existing barriers. Cognitive radio was first introduced by Mitola and Maguire (1999; 2000) for the

Cognitive Radio and Dynamic Spectrum Access   L. Berlemann and S. Mangold
© 2009 John Wiley & Sons, Ltd

flexible and efficient usage of spectrum. Cognitive radio can be regarded as the necessary step from a flexible physical layer up to a flexible system as a whole. The term 'cognitive radio' is derived from 'cognition'. Often, cognition is referred to as information processing, involving learning and knowledge.

Cognitive radio is therefore a self-aware communication system that efficiently uses spectrum in an intelligent way in applying dynamic spectrum access. It autonomously coordinates the usage of spectrum by identifying unused radio spectrum on the basis of observing spectrum usage. The consideration of spectrum as being unused and its usage involves regulation, as this spectrum might be originally assigned to a licensed communication system. This was introduced as 'vertical spectrum sharing' earlier in this book. Besides cognition in radio resource management, cognition in services and applications is provided by cognitive radios to enable transparency to the user. The mental processes of a cognitive radio based on the cognition circle from Mitola (2000) are shown in Figure 7.1.

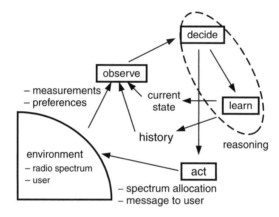

Figure 7.1 Mental processes of a cognitive radio based on the cognition cycle from Mitola (2000)

Cognition is illustrated by the example of flexible radio spectrum usage and consideration of user preferences. In observing the environment, a cognitive radio decides about its action. An initial switching on may lead to an immediate action, while usual operation implies a decision taking based on learning from observation history together with the actual state of the environment.

The FCC has identified (FCC, 2005c) the following features that cognitive radios may incorporate to allow a more efficient, flexible spectrum use:

- **Frequency agility** – The radio is able to change its operating frequency to optimize its use in adapting to the environment.
- **Dynamic Frequency Selection (DFS)** – The radio senses signals from nearby transmitters to choose an optimal operation environment.
- **Adaptive modulation** – The transmission characteristics and waveforms can be reconfigured to exploit all opportunities for the usage of spectrum.
- **Transmit Power Control (TPC)** – The transmission power is adapted to full power limits when necessary on the one hand and to lower levels on the other hand to allow greater sharing and reuse of spectrum.

- **Location Awareness** – The radio is able to determine its location and the location of other devices operating in the same spectrum to optimize transmission parameters for increasing spectrum reuse.
- **Negotiated Use** – The cognitive radio may have algorithms enabling the sharing of spectrum in terms of prearranged agreements between a licensee and a third party or on an *ad-hoc*/real-time basis, see Section 3.1 and thereafter.

Strictly following this definition, modern wireless LANs could already be regarded as cognitive radios, as IEEE 802.11 devices operate with a listen-before-talk spectrum access and with dynamically changing frequencies and transmission power.

Cognitive radios may also be referred to as 'spectrum agile radios' (Mangold *et al.*, 2004a; 2005a) to indicate emphasis on dynamic spectrum access. Mangold *et al.* (2004a) focus therefore on IEEE 802.11k for radio resource measurements as an approach to facilitating the development of spectrum agile radios, while Mangold *et al.* (2005a) introduce spectrum agile radios as a group of value oriented machines. Basic concepts from social science are used to classify the social action of independent decision makers and are applied to define system strategy rules. We discuss this later in the context of spectrum etiquette in Section 7.4.2.

A similar understanding of cognitive radios is summarized in the definition of cognitive radio from Haykin (2005):

> 'Cognitive radio is an intelligent wireless communication system that is aware of its surrounding environment (i.e., outside world), and uses the methodology of understanding-by-building to learn from the environment and adapt its internal states to statistical variations in the incoming RF stimuli by making corresponding changes in certain operating parameters (e.g., transmit power; carrier-frequency, and modulation strategy) in real-time, with two primary objectives in mind: (i) highly reliable communication whenever and wherever needed, and (ii) efficient utilization of the radio spectrum.'

Highly reliable communication includes in this context the aspect of QoS. Cognitive radios have to share spectrum and therefore require means of mutually coordinating their spectrum utilization to enable the support of QoS in such distributed environments.

Cognitive radio networks are a logical generalization of cognitive radios. The extension of the focus from the individual cognitive radio to a cognitive radio network aims especially at joint improvement of spectrum utilization through spectrum reuse and common spectrum sensing. Policy-based reasoning, as described in Section 5.3, enables operator-specific, cognitive radio network deployments. Policies enable the distributed coordination necessary to realize self-configuration. They define the characteristics for the consideration of a link as a relay link of the wireless infrastructure and take this into account in spectrum usage. Policies specify the restrictions to spectrum usage in terms of bandwidth and time of spectrum access that are required for the relay links to enable a QoS guarantee. From the operator perspective, policies can be used to define behaviors of cognitive radios in case of relaying requests from radios belonging to the same operator or others.

A comparable approach is introduced by Buddhikot *et al.* (2005) with the Dynamic Intelligent Management of Spectrum for Ubiquitous Mobile-access Networks (DIMSUMnet) architecture that enables a coordinated, real-time, access to spectrum. The focus of the DIMSUMnet is limited to improvement of access to spectrum in time, frequency and space domain – different to an opportunistic spectrum usage as discussed at many places in this book. The authors also see the advantages of a relay-based, systems enabling, coordinated spectrum reuse and introduce therefore a DIMSUM-RelayCluster architecture.

7.2 Cognitive Radios Can Gain from Delay-tolerant Software Radio

We support the hypothesis that true cognitive radio will gain from software (defined) radio and reconfigurabiliy if true software definition can be realized.

The basic intent of software-based radios was always to shift the hardware-oriented, application-specific, implementation of communication devices to flexible software applications performing communication functions on a common computing platform. If the communication functions of a transceiver are completely realized as programs operating on a suitable processor, this transceiver is referred to as Software Radio (SR) (Mitola, 1995). The digitalization is realized directly after radio wave reception at the antenna and all the signal processing is done by software. Software Defined Radios (SDRs) (Tuttlebee, 2002a, b) are a first step in the evolution to SRs and thus more practicable, as the analog signals are processed after a suitable band filter is selected. SDRs are factual reality when taking the recent convergence of software and digital radio into account. Though the SR concept is related to all layers of the protocol stack, many research efforts concentrated in the past on the PHYsical layer (PHY). In the last five years the scope was extended to the complete software of a device used for communication in introducing new concepts, which are all based on SDRs as introduced in the following sections. A communication system of SDRs can have the following characteristics:

- A multi-band system operating in multiple frequency bands, such as for example in the ISM bands at 2.4 and 5 GHz.
- A multi-standard system supporting more than one radio standard of the same protocol family (e.g., IEEE 802.11a/b/g) or of different radio access technologies (UMTS, GPRS and IEEE 802.11b).
- A multi-channel system enabling simultaneous transmission and/or reception on multiple channels.

The multi-mode systems that are currently the focus of many research initiatives combine these three characteristics. The SDRforum expanded its understanding of SDR to the same extent as described by the reconfigurable radio concept below (SDRforum, 2005).

An initial step towards a semiformal software specification is presented by Schinnenburg *et al.* (2005) as a framework for building reconfigurable protocol stacks. A high degree of reconfigurability is achieved by composing complex behavior of a communication system using functional units. This work has its origin in Siebert and Walke (2001) and Berlemann *et al.* (2003a) where a design technique for enhancing existing and developing future 3G and 4G protocol stacks is proposed and is referred to as a generic protocol stack.

Software (defined) radio has been developed in the hope of improved flexibility in spectrum access. For example, traceable decision making, certified mobile computing algorithms, and cognitive radio, are today sufficiently understood. In contrast, commonly used radio systems still require real-time operation and are for this reason implemented by hardware. They are not software defined, not cognitive, and usually cannot be modified after manufacturing.

Today's layer-2 protocols and layer-1 access schemes do not allow a nonreal time and low cost software radio approach. This is not desirable and should be addressed in the future.

Too many time-critical steps are involved when radio devices communicate wirelessly. For example, radio waves are modulated at predefined and exact symbol time durations that must not alter throughout data exchange. Protocols work with time-critical acknowledgements and announcements, which require operational alignment in time between devices, very often in the range of microseconds. Such systems cannot be realized for example in a low cost JAVA virtual machine with its unpredictable garbage collection, or run on software of a common computer laptop.

Protocols and access schemes that manage unexpected delays and survive interruptions at any random interval are needed. Computers that do not necessarily operate with real-time kernels could provide the needed resources to realize flexible spectrum access. Such a delay-tolerant approach would open the path towards commercial exploitation of cognitive radios.

7.3 DARPA XG Provides Implementation Guidelines, Including the Access Protocol

We are convinced from what is known about the outcome of the DARPA XG program, that the semantic reasoning concepts proposed by XG will provide major benefits for use in future standardization of wireless communications networks.

True cognition requires semantic concepts that can be implemented with ontology representations. Sets of ontology representations can be used to describe knowledge about the radio domain and further to describe algorithms about how to utilize radio resources, thus enabling cognition. With the objective of opening radio spectrum with the help of cognitive radio technology, and the concepts that we may adopt from social science and game theory, the new engineering methods by XG for developing the radio systems of the future should be explored in more detail. Cognitive radios have to make decisions that are traceable, and when decisions are made, a cognitive radio may be required to formulate its objectives (goal of reasoning), and how it came to the final decision. The following sections are based on the initial ideas on policies from Section 5.3.

7.3.1 Traceable Decision Making

Traceability is needed for two reasons. The first one is the regulatory accreditation of the spectrum access algorithms that are used by cognitive radios. Radio regulations will be described not in texts to be read by engineers, but in a way such that devices will be able to make decisions on their own. The way decisions are made will be disclosed to regulators, and in this way, the regulators will remain in control of which system operates in which spectrum. The second reason for traceable decision making is that algorithms can be patented. With traceable algorithms, innovative ideas that find their way into the silicon chips of cognitive radios can be disclosed. This increases the value of patents since patent infringement can now be detected. With patent infringement detection, chip manufacturers may experience reasonable incentives to build such devices and protect their innovations. This may stimulate progress towards commercial exploitation of cognitive radio and dynamic spectrum access, in the case of a reasoning approach similar to that of XG being generally applied.

To realize this vision, knowledge of wireless communication should be described in a machine-understandable way, and engineering machine-understandable radio semantics using ontologies is one of the main approaches that XG tries to solve.

7.3.2 Machine-understandable Radio Semantics

A taxonomy classifies information entities so that they can be browsed or navigated for information. A taxonomy is a description in which information entities are related to each other. An example of taxonomy is 'unused spectrum is a spectrum opportunity.' Such expressions can be used to express semantic content with relationships such as for example 'synonym for,' or 'associated with.'

An ontology extends the concept of taxonomy, and is a collection of classes and properties that are combined with each other. Ontology extends the concept of taxonomy towards a logical theory, and is used to capture the meaning of a subject area, such as the radio domain or an area of knowledge that corresponds to what engineers know about this domain. Ontology makes radio semantics domain knowledge usable by way of the following concepts:

- classes and instances in the knowledge domains of interest;
- relationships;
- attributes;
- (propositional) functions and processes;
- constraints and rules.

Ontologies encode the description of knowledge that enables a cognitive radio to use that description. Ontology tools such as Stanford's Java Theorem Prover (JTP) can perform automated reasoning about the knowledge. One established language for representing ontology is the Web Ontology Language (OWL) (McGuinness and Harmelen, 2003; Smith *et al.*, 2003). OWL is a rich language based on XML, that allows not only first-order logic, but also higher-order, class-based, reasoning (Baader *et al.*, 2003).

Ontologies represented as logical theories can be directly interpreted by a cognitive radio. Logical theories are built on axioms and inference rules. Axioms are statements that are declared as true. Inference rules are rules that, given assumptions, provide valid conclusions. Axioms together with inference rules can be used to prove theorems about the knowledge domain. The set of axioms, inference rules, and theorems together constitute the logical theory to express radio semantics.

With an ontology set describing a domain of knowledge, logic such as the simple first order predicate logic enables cognitive radios to make decisions. A predicate (propositional function) is used to make a statement. If it is said 'radio resource A is an opportunity,' then the predicate 'is an opportunity' is applied to a radio resource. We also might say that we have predicated 'being opportunity' of radio resource A, or attributed 'opportunity' to radio resource A.

A predicate applies to an instance (in our example, radio resource A) or a class. It hence generates a potentially valid and new proposition (for example 'A is an opportunity').

7.3.2.1 The Wi-Fi MAC Protocol as Policy

We now discuss how to use a policy language to specify a MAC protocol. As example, we have used the IEEE 802.11e protocol for QoS support in WLAN.

The Enhanced Distributed Channel Access (EDCA) realizes the contention-based access of the Wireless LAN IEEE 802.11e under decentralized operation (IEEE, 2005c). IEEE 802.11e provides differentiated QoS. To support QoS, the EDCA introduces four Access Categories (ACs). Each AC relates to one corresponding backoff entity. The four backoff entities of a WLAN station operate in parallel and realize contention-based access. The four ACs of 802.11e are AC_BK ('background'), AC_BE ('best effort'), AC_VI ('video') and AC_VO ('voice'). They are derived from the user priorities from Annex H.2 of IEEE 802.1D (IEEE, 1998). The prioritization between the four backoff entities is realized by different AC-specific parameters, i.e. the EDCA parameters sets. These EDCA parameter sets modify the backoff process with individual interframe spaces and contention window sizes per AC introducing a probability-based prioritization as explained below (Figure 7.2).

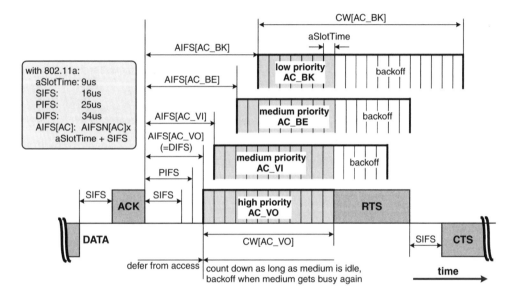

Figure 7.2   EDCA timing diagram of the four backoff entities defined in 802.11e with different AIFSs and contention window sizes

The EDCA parameters of each backoff entity are defined by the HC and may be adapted over time. Default values for the EDCA parameters are given in IEEE (2005c) and Table 7.1. Within a station, each backoff entity individually contends for obtaining a spectrum opportunity.

A backoff entity starts decreasing its backoff counter after detecting that a channel is idle for an Arbitration Interframe Space (AIFS). The AIFS has a duration of at least DCF Interframe Space (DIFS) and depends on the corresponding AC of the four ACs of 802.11e. To express this dependency, it is denoted therefore in the following as *AIFS[AC]*. The Short Interframe Space (SIFS) is the shortest interframe space of 802.11. It is used between the frames of the RTS/CTS/DATA/ACK sequence. The PCF Interframe Space (PIFS) is used by the PCF to gain access to the radio channel. The Arbitration Interframe Space Number (AIFSN) is defined per AC according to (IEEE, 2005c) and enlarges *AIFS[AC]*. A small *AIFS[AC]* implies a high access priority. The earliest channel access time after an idle channel, i.e., the shortest value

Table 7.1 Default values of EDCA parameters based on IEEE (2005c). The star indicates dependency on a physical layer, here 802.11a

AC	CWmin	CWmax	AIFSN	AIFS
Legacy	15	1023	2	34 us
AC_BK	15	1023	7	79 us
AC_BE	15	1023	3	43 us
AC_VI	7	15	2	34 us
AC_VO	3	7	2	34 us

of $AIFS[AC\_VO]$ = DIFS is similar to the legacy DCF of 802.11. Prioritization is reached in this case through different values of the following contention window.

The Contention Window (CW) of the backoff process is used in 802.11e to introduce priorities further. For the legacy 802.11a PHY, the minimum and maximum values are given by $Cwmin = 15$ and $Cwmax = 1023$. A small $Cwmin[AC]$ leads to a high access priority.

7.3.2.2 Policy Description

The capabilities, parameters and processes that are required to enable a cognitive radio to operate according to the EDCA of 802.11e are specified in Policy 7.1. The way an EDCA spectrum opportunity is obtained is described in Policy 7.2. Opportunity (OppDesc) and usage descriptions (UseDesc) are combined and define policies (PolicyRule) for spectrum

Policy 7.1 Device capabilities, parameters and processes of the EDCA expressed in the shorthand notation of the DARPA XG policy language

```
1  (DeviceCap (id 802.11EDCA_Profile)
      (hasPolicyDefinedParams
        /* constant parameters */
        CWmax CWmin AIFSN AIFS
        /* variable parameters */
        BackoffCounter CW QSRC QLRC
        /* parameters bound by 802.11 device */
        aSlotTime aSIFSTime
        dot11ShortRetryLimit dot11LongRetryLimit)
      (hasPolicyDefinedBehaviors
        random /* draws random integer value */
        SenseIdleChannelDuration
        InitiateFrameSequence
        SenseSlot DiscardAttempt))
2  (Process (id random)
      (input lower_border upper_border)
      (output random_value))
```

Policy 7.1   (*continued*)

---

3   (TimeDuration (id aSlotTime)
      (boundBy Device) (unit msec))
     (TimeDuration (id aSIFSTime)
      (boundBy Device) (unit msec))
     (RetryCnt (id dot11ShortRetryLimit)
      (boundBy Device) (unit NONE))
     (RetryCnt (id dot11LongRetryLimit)
      (boundBy Device) (unit NONE))

4   (Integer (id AIFSN)
      (magnitude 2) (unit NONE)) /* AC_VI*/
     (CWsize (id CWmin))
      (magnitude 7) (unit NONE)) /* AC_VI*/
     (CWsize (id CWmax))
      (magnitude 15) (unit NONE)) /* AC_VI*/

5   (TimeDuration (id AIFS)
      (magnitude
      (xgx '(+(* AIFSN aSlotTime) aSIFSTime)'))
      (unit msec))

---

usage. These policies are aggregated in a group of policies (PolicyGrp). This group is specified in Policy 7.3 and represents all rules of the EDCA. The reasons for invoking the backoff procedure are specified as spectrum opportunity descriptions in Policy 7.4. The manipulation of the EDCA parameters set in the backoff procedure according to the standard are described as usage descriptions in Policy 7.5. We assume that the EDCA's policies are repeatedly processed by a cognitive radio for each idle time slot. Additionally, the policies are processed at the end of a (failed or successful) frame transmission sequence.

Policy 7.2 Procedure for obtaining an EDCA spectrum opportunity expressed in the shorthand notation of the DARPA XG policy language

---

1   (PolicyRule (id TransmitFrameSequence)
      (selDesc 802.11_5GHz_US) (deny FALSE)
      (oppDesc Idle1) (useDesc InitTrans))

2   (OppDesc (id Idle1) (xgx '(and
      (invoke SenseIdleChannelDuration
      TimeDuration IdleChannelDuration)
      (>= IdleChannelDuration AIFS)
      (eq FrameAvailable BoolTrue)
      (= BackoffCounter 0)
      (eq HigherPriorTransmit BoolFalse))'))

3   (UseDesc (id InitTrans)
      (xgx '(invoke InitiateFrameSequence)'))

(*continued overleaf*)

Policy 7.2   *(continued)*

4 (PolicyRule (id DecreaseBackoffCounter)
   (selDesc 802.11_5GHz_US) (deny FALSE)
   (oppDesc Idle2)(useDesc DecBackoff))

5 (OppDesc (id Idle2) (xgx '(and
   (invoke SenseIdleChannelDuration
   TimeDuration IdleChannelDuration)
   (>= IdleChannelDuration AIFS)
   (< BackoffCounter 0)'))

6 (UseDesc (id DecBackoff) (xgx '(:=
   BackoffCounter (- BackoffCounter 1))'))

7 (PolicyRule (id InternalCollision)
   (selDesc 802.11_5GHz_US) (deny FALSE)
   (oppDesc Idle3) (useDesc Backoff2))

8 (OppDesc (id Idle3) (xgx '(and
   (invoke SenseIdleChannelDuration
   TimeDuration IdleChannelDuration)
   (>= IdleChannelDuration AIFS)
   (eq FrameAvailable BoolTrue)
   (= BackoffCounter 0)
   (eq HigherPriorityTransmit BoolTrue))'))

Policy 7.3   All policies specifying an operation according to the EDCA are gathered in
a policy group

1   (PolicyGrp (id 802.11EDCA)
      (equalPrecedence TRUE)
      (polMembers
      TransmitFrameSequence
      DecreaseBackoffCounter
      InternalCollision
      TransSucc1 ... TransSucc7
      BusyChannel1 BusyChannel2
      TransFail1 ... TransFail5))

2   (PolicyRule (id TransSucc1)
      (selDesc 802.11_5GHz_US)
      (deny FALSE) (oppDesc
      Success1)
      (useDesc Success))

      . . .

      (PolicyRule (id TransSucc1)
      (selDesc 802.11_5GHz_US)
      (deny FALSE) (oppDesc
      Success7)
      (useDesc Success))

Policy 7.3 *(continued)*

---

3   (PolicyRule (id BusyChannel1)
       (selDesc 802.11_5GHz_US)
       (deny FALSE) (oppDesc Busy1)
       (useDesc Backoff1))
     (PolicyRule (id BusyChannel2)
       (selDesc 802.11_5GHz_US)
       (deny FALSE) (oppDesc Busy2)
       (useDesc Backoff1))

4   (PolicyRule (id TransFail1)
       (selDesc 802.11_5GHz_US)
       (deny FALSE) (oppDesc Fail1)
       (useDesc Backoff2))

     ...

     (PolicyRule (id TransFail5)
       (selDesc 802.11_5GHz_US)
       (deny FALSE) (oppDesc Fail5)
       (useDesc Backoff2))

Policy 7.4   Reasons for invoking the EDCA backoff procedure expressed as spectrum opportunities in the shorthand notation of the DARPA XG policy language

---

1   (OppDesc (id Success1) (xgx '(and
       (invoke SenseSlot
       SlotStateType SlotState)
       (eq SlotState CTSonRTS)'))

2   (OppDesc (id Success2) (xgx '(and
       (invoke SenseSlot
       SlotStateType SlotState)
       (or (eq SlotState MPDU)
       (eq SlotState BlockAck))'))
     ...

3   (OppDesc (id Busy1) (xgx '(and
       (eq FrameAvailable BoolTrue)
       (= BackoffCounter 0)
       (invoke SenseSlot
       SlotStateType SlotState)
       (eq SlotState PhysicalCS)'))

4   (OppDesc (id Busy2) (xgx '(and
       (eq FrameAvailable BoolTrue)
       (= BackoffCounter 0)
       (invoke SenseSlot
       SlotStateType SlotState)
       (eq SlotState VirtualCS)'))

(continued overleaf)

Policy 7.4   (*continued*)

5 (OppDesc (id Fail1) (xgx '(and
      (invoke SenseSlot
      SlotStateType SlotState)
      (eq SlotState failCTSonRTS)')))

6 (OppDesc (id Fail2) (xgx '(and
      (invoke SenseSlot
      SlotStateType SlotState)
      (eq SlotState failACKonMPDU)')))

      ...

Policy 7.5   Parameter manipulation of the EDCA backoff procedure expressed in shorthand notation of the DARPA XG policy language

1 (UseDesc (id Success) (xgx '(and
      (:= CW CWmin)
      (:= BackoffCounter random(0,CW))
      (:= QSRC 0) (:= QLRC 0))')))

2 (UseDesc (id Backoff1) (xgx '(and
      (:= CW CW) /* CW remains fixed */
      (:= QSRC (+ QSRC 1)) /*QSRC=QSRC+1*/
      (:= QLRC (+ QLRC 1)) /*QLRC=QLRC+1*/
      (:= BackoffCounter
      random(0,CW)))')))

3 (UseDesc (id Backoff2) (xgx '(and

4 (if (or (= QSRC dot11ShortRetryLimit)
      (= QLRC dot11LongRetryLimit))
      (and (:= CW CWmin)
      (:= QSRC 0) (:= QLRC 0)
      (invoke DiscardAttempt)))

5 (if (or (< QSRC dot11ShortRetryLimit)
      (< QLRC dot11LongRetryLimit))
      (and
      (if (< CW CWmax)
      (:= CW (-1 (* 2 (+ CW 1)))))
      /* CW=(CW+1)·2-1) */
      (if (= CW CWmax) (:= CW CW))
      (:= QSRC (+ QSRC 1))
      (:= QLRC (+ QLRC 1))))

6 (:= BackoffCounter random(0,CW)))')))

The slotting in the time domain is introduced by *aSlotTime,* which depends on PHY mode used by the 802.11e MAC. In case of 802.11a, a time slot has, for example, a duration of 9 μs. The PHY mode dependent parameters of 802.11 are provided by the cognitive radio (boundBy Device) to enable the processing of the EDCA policies. These parameters are specified in Policy 7.1 together with the processes required for executing the EDCA policies. The EDCA parameter set of an AC (here AC_VI) is defined in Policy 7.1, line 4 according to Table 7.1. A backoff entity with a different AC would here assign different values to the parameters of the backoff procedure.

7.3.2.3 Obtaining a Spectrum Opportunity

Before attempting a transmission, a backoff entity decreases its backoff counter when detecting an idle channel for the duration of *AIFS[AC]*. The following description of the ECDA procedures for obtaining a spectrum opportunity is based on IEEE (2005c), there called transmission opportunity.

Each backoff entity maintains a backoff counter, which specifies the number of backoff slots an entity waits before initiating a transmission. The duration *AIFS[AC]* is defined in accordance with:

$$AIFS[AC] = SIFS + AIFSN[AC] \cdot aSlotTime.$$

AIFS[AC] is specified in Policy 7.1, line 5. The attempt to obtain a spectrum opportunity is determined according to the following conventions.

The backoff entity of an AC performs on specific slot boundaries, defined by *aSlotTime,* exactly one of the functions below. The conditions for performing one of these functions are reflected in the opportunity descriptions of unused spectrum (here a frequency channel). Usage and opportunity description form together the EDCA policies for:

- **Obtaining a spectrum opportunity**
  (TransmitFrameSequence, Policy 7.2, lines 1–3): Initiate a frame exchange sequence (useDesc InitTrans) if (i) there is a frame available for transmission, (ii) the backoff counter has reached zero, and (iii) no internal backoff entity with higher priority is scheduled to initiate a transmission. These three conditions form, together with the channel idle time of AIFS, the spectrum opportunity description Idle1.
- **Decrementing the backoff counter**
  (DecreaseBackoffCounter, Policy 7.2 lines 4–6): The backoff counter is decremented (useDesc DecBackoff) if it has a nonzero value. This leads to opportunity description Idle2.
- **Invoking the backoff procedure**
  (useDesc Backoff2) because of an internal collision (InternalCollision, Policy 7.2, lines 7–8) if (i) there is a frame available for transmission, (ii) the backoff counter has reached zero, and (iii) an internal backoff entity with higher priority is scheduled to initiate a transmission (oppDesc Idle3). This rule can also be found below in the EDCA backoff procedure as (oppDesc Fail4).
- **Doing nothing**. This function requires no specification as policy.

The specific slot boundaries, at which one of these operations is performed, essentially depend on the point of time after which the channel is regarded as being idle. These boundaries are defined for each backoff entity in IEEE (2005c) by introducing modifications to the $AIFS[AC]$ from above. We neglect these modifications in the following for the sake of simplicity.

7.3.2.4 EDCA Backoff Procedure

The backoff procedure of an AC is invoked in the case of a transmission failure or of a virtual collision due to an internal transmission attempt by multiple ACs. Each backoff entity of the EDCA has a state variable $CW[AC]$ that represents the current size of the contention window of the backoff procedure. $CW[AC]$ has an initial value of $CWmin[AC]$. The size of the contention window $CW_i[AC]$ in backoff stage i is defined therefore as:

$$CW_i[AC] = min\left[2^i\left(CWmin[AC]+1\right)-1, CWmax[AC]\right].$$

This definition is specified in Policy 7.5, line 5.

In the case of a successful frame transmission $CW[AC]$ is reset to $CWmin[AC]$ (useDesc_Success). A successful transmission is indicated by:

- reception of a CTS in response to an RTS (oppDesc Success1, Policy 7.4, line 1);
- reception of a unicast MPDU or BlockAck (oppDesc Success2, Policy 7.4, line 2);
- reception of a BlockAck in response to a BlockAckReq (oppDesc Success3, not specified here);
- reception of an ACK in response to a BlockAckReq (oppDesc Success4, not specified here);
- transmittion of a multicast frame with a 'no acknowledgement' policy (oppDesc Success5, not specified here);
- transmitting of a frame with a 'no acknowledgement' policy (oppDesc Success6, not specified here).

The backoff procedure of a backoff entity is invoked when:

- (i) A frame is intended to be transmitted, (ii) the backoff counter has reached a value of zero, and (iii) the medium is busy. This may be indicated by either a physical (oppDesc Busy1, Policy 7.4, line 3) or virtual (oppDesc Busy2, Policy 7.4, line 4) carrier sense. In this case, the backoff procedure is invoked and the value of $CW[AC]$ remains unchanged (useDesc Backoff1, Policy 7.5, line 2).
- The final transmission of a transmission opportunity holder during its transmission opportunity is successful (OppDesc Success7, not specified here). The value of $CW[AC]$ is reset to $CWmin[AC]$ (useDesc Success, Policy 7.4, line 1).
- A frame transmission fails. This is indicated by a failure to receive a CTS in response on an RTS (oppDesc Fail1, Policy 7.4, line 5), a failure in receiving an ACK that is expected on a unicast MPDU (oppDesc Fail2, Policy 7.4, line 6), a failure to receive a BlockAck in response to a BlockAckReq (oppDesc Fail3, not specified here) or a

failure to receive an ACK in response to a BlockAckReq (oppDesc Fail4, not specified here).
- The transmission attempt of an AC collides internally with a higher priority AC (oppDesc Fail5, not specified here).

In the case of a frame transmission failure the value of $CW[AC]$ is updated as described in the following (useDesc Backoff2, Policy 7.5, lines 3–6) before invoking the backoff procedure:

- In the case that $QSRC[AC]$ or $QLRC[AC]$ has reached dot11ShortRetryLimit or dot11 LongRetryLimit respectively, $CW[AC]$ is reset to $CWmin[AC]$ and the transmission attempt is discarded (Policy 7.5, line 4). Otherwise, $CW[AC]$ is set to $(CW[AC]+1) \cdot 2 - 1$ when $CW[AC] < CWmax[AC]$ or $CW[AC]$ remains unchanged if $CW[AC] = CWmax[AC]$. For the rest of the retransmission attempts the size of the contention window is not changed (Policy 7.5, line 5).

After setting the contention window size $CW[AC]$, the backoff procedure sets the backoff counter to a randomly chosen integer value with a uniform distribution over the interval $[0,CW[AC]]$ (Policy 7.5, line 6).

7.4 Spectrum Etiquette May Stimulate Cognitive Behavior

We support the proposition that spectrum etiquette should be part of future radio systems to stimulate cognitive behavior.

Cognitive radios often coordinate radio resources autonomously while operating. With this approach, fair and efficient resource sharing between radio systems is difficult to achieve. Horizontal spectrum sharing is a challenge for cognitive radio as well as for traditional radio systems operating in unlicensed frequency bands. It is the objective of spectrum etiquette rules to provide incentives that will motivate radio systems to allocate spectrum more efficiently.

7.4.1 What is Spectrum Etiquette?

A spectrum etiquette is a set of rules for radio resource management to be followed by cognitive radio systems that share the radio spectrum. Spectrum etiquette often helps to establish fair access to the available radio resources, in addition to a more efficient usage of radio spectrum. Spectrum etiquette rules are typically based on actions like dynamic channel selection, transmission power control, adaptive duty cycles, and carrier sensing (listen-before-talk).

A spectrum etiquette defines the rules for the behavior of radio systems mainly in order to achieve two goals. Firstly, if all radio systems follow the spectrum etiquette, fairness in access to the shared radio resources is maintained, and secondly, the frequency band is more efficiently used. A spectrum etiquette is independent of any radio transmission scheme and protocol; it does not define a protocol and is not restricted to one radio standard. Further, a spectrum etiquette is not an algorithm that describes the entire radio resource management of all radio systems. Each radio system can apply its own algorithms within the constraints of the spectrum etiquette. The spectrum etiquette provides a framework for behaviors, which may restrict the degrees of freedom in radio resource management of the individual radio systems, but leaves room for innovations and differentiations between devices from different vendors.

In order to motivate manufacturers to implement spectrum etiquette rules, convincing sets of rules are needed that provide incentives for radio systems to operate spectrum efficiently, for the benefit of all radio systems.

7.4.2 Value Orientation

Spectrum sharing between different radio systems can be understood as a scenario forming a society of independent decision-makers (Mangold *et al.*, 2005a; 2006a). Therefore, basic concepts to classify social action that are taken from social science can be applied to define system strategy rules. The rules represent algorithms for decision-making entities that reside in the cognitive radio systems.

A contemporary society is a group of socially acting individuals (i.e., 'actors') where each individual acts according to classified motivations. An action is for example the selection of radio transmission parameters such as transmission powers and the frequency of operation. An actor's interests can be expressed through their application requirements (for example, the throughput in Mb/s that an application would require), as it is typically applied when games are used to analyze communication protocols.

However, we are not only interested in the economic self-interests of actors, but also the actor's value orientations. Value orientation is a concept in contemporary societies with social awareness. Sociological models known from Max Weber's early work may be applied to build a framework for a society of machines. These models allow the design of new types of radio systems that are not only self-aware, but also socially aware. This approach for autonomous spectrum sharing supports our goal of achieving technology-based radio regulation, with cognitive radios that manage and regulate spectrum without human interaction. By introducing value orientation, cognitive radios can be designed so that they are capable of coordinating their spectrum usage. Whereas until today, cognitive radio systems were thought of as pure technocratic decision makers, value-oriented cognitive radios will support the introduction of true cognition, for example for the purpose of higher efficiency in spectrum usage.

Cognitive radios that share the radio spectrum can be interpreted as forming a virtual society. This virtual society faces many challenges that have been already analyzed for real-life societies, such as for example interacting nations, or groups of animals such as ant colonies. The theory of games is an efficient means of analyzing such societies. Cognitive radios will be required to act as environment and interference aware when operating with shared radio resources. They have to operate with high flexibility, by also considering the implications of their decisions on other radio systems. It may be helpful to extend the traditional way that cognitive radios are defined by introducing social awareness. Cognitive radios that are aware of their society, i.e. aware of the existence and demands of other radio systems, can benefit by supporting not only their own interests, but also the interests of the other radio devices.

Cognitive radios make decisions about what action to take; hence the reference to 'actor'. There is a predefined set of valid actions that an actor can take; this set spans the so-called action space. According to Weber, an action is usually associated with a subjective meaning. Two classes of action exist: social action and economic action. An action is social, if 'its subjective meaning takes account of the behavior of others and is thereby oriented in its course.' An action that intends to improve the actor's outcome is an economic action. However, the subjective orientation of an actor is what we are particularly interested in when discussing the concept of social action.

Weber's classification of social action identifies four types:

- technocratic social action ('zweckrational');
- value-oriented social action ('wertrational');
- affective social action (based on emotional state);
- traditional social action (guided by custom).

It is helpful to interpret the existing way of radio regulation and spectrum sharing as technocratic social action: whatever other individuals observe, each individual attempts to optimize his own spectrum usage only. This often leads into chaotic and unpredictable scenarios of spectrum usage as if there had been no regulation: the purely technocratic social actions of today's radio systems leverage the existence of radio regulation. However, we are interested in opening the spectrum for free usage, with a minimum amount of radio regulatory constraints. Therefore, motivated by the findings of Max Weber, it may be helpful to apply value-oriented social action – the same social concepts that allow human beings to live without dominant regulation of their daily interactions. For the spectrum-sharing problem, the decision-making processes in cognitive radios are then modified to support the social concepts of value orientation, and hence implement voluntary rules.

7.5 Network Operators May Assist Dynamic Spectrum Access

'Operator-assistance' (Mangold *et al.*, 2006b, c) introduces centralized means for coordinating spectrum access to cognitive radio. Such a centralized approach has the advantage of facilitating the regulator's control of spectrum usage, and allowing them to direct how the spectrum is used.

We have already outlined the basic concept of our approach on operator-assistance earlier in this book (see, for example, Section 3.3.3). To protect licensed radio systems, for example in vertical spectrum sharing, while gaining from the flexibility of decentralized approaches, radio systems owned by operators can indeed efficiently assist cognitive radios in identifying underutilized spectrum. In this way, dynamic coordination of spectrum usage is facilitated with the help of a radio network that has clearly defined site locations, and a trusted operator.

Established incumbent cellular network operators usually have a reasonable relationship with radio regulators and hence may get permission to assist cognitive radios to make the decision about when spectrum is declared underutilized.

7.6 Business Opportunities

The introduction of cognitive radio may provide new opportunities for the business models of wireless communication (Berlemann *et al.*, 2006c).

We summarize some basic thoughts on the implications of cognitive radio for incumbent operators in an analysis illustrated in Table 7.2. The main advantage of cognitive radio for operators is the elimination of the lengthy and expensive licensing process. Operators have therefore the chance to realize a multitude of new services.

The complexity of devices, primary radio protection and unsatisfying QoS support are technical problems that are currently being discussed in the research domain.

Table 7.2   Analysis of cognitive radio from incumbent operator's (today's license holders) perspective

Strengths	*Weaknesses*
• Improved spectrum efficiency	• Protection of licensed radio systems against in interference may not be reliable
• Eliminated need for expensive and lengthy spectrum licensing process	
• More flexible spectrum assignment	• May not be sufficient for services that require restrictive QoS
• Less technology-dependent access to spectrum	• Low acceptance by license owners and regulators
• Seriously considered by US regulators → potential leverage effect	• Increased complexity of devices
	• Additional infrastructure may be required
	• Multiple value chain participants
Opportunities	*Threats*
• New types of services like operator assisted dynamic spectrum assignment (to protect licensed services and to enable QoS) may provide new revenue stream	• Thresholds for entering market of commercial wireless communication are lowered
• Spectrum trading (lease, reselling, etc.) and interoperator spectrum sharing	• New upcoming competitors
	• Cognitive radio devices may be allowed to operate in incumbent operator's spectrum
• Potential for new very high bit rate systems (Gb/s)	• Operators may lose competitiveness if trend is missed

A major threat for incumbent network operators is the breakdown of market entry thresholds. The highly paid exclusiveness of spectrum access to certain frequency bands may be lost and the investment costs for building up a radio network will not consider spectrum license fees anymore. New competitors are then able easily to enter the market and can attack incumbent operators and licenses will in the long term expire without being renewed by the regulator. A timely introduction of operator-assisted cognitive radio services and coordinated secondary spectrum access are a promising answer to dealing with this threat.

With cognitive radio on the horizon, spectrum assignment and its licensing will become more flexible and dynamic. Secondary usage of spectrum in vertical spectrum sharing scenarios can be identified as a major pillar in the current liberalization of spectrum regulation.

The general options for a new way of spectrum regulation, for spectrum sharing between primary and secondary radio systems, are summarized in Table 7.3. Example application scenarios are mentioned that consider especially the sharing models in long-range communication that effect cellular network operators. Such operators may be both primary and secondary users of spectrum. Spectrum sharing scenarios with coordination between primary and secondary systems allow temporary QoS guarantees and are therefore an option for operators that are secondary users of spectrum.

As we know, one of the main problems of the distributed nature of operating cognitive radios with opportunistic spectrum access is the mitigation of interference. Interference occurs

Table 7.3 Spectrum regulation options for spectrum sharing between primary and secondary radio systems as a refinement of Peha (2005)

	Secondary radio system not licensed	Secondary radio system licensed
No coordination between primary and secondary	Unlicensed spectrum sharing • primary: TV broadcaster • secondary: opportunistic access of cognitive radios without QoS guarantee	Licensed secondary spectrum access • primary: TV broadcaster • secondary: cellular network operator with exclusive access, using guard bands etc.
Coordination between primary and secondary	Spectrum trading at secondary markets • primary: cellular network operator I • secondary: cellular network operator II or cognitive radios with temporary QoS guarantees	Interruptible spectrum access • primary: public safety or military services • secondary: cellular network operator with temporary QoS guarantee

between operators, dissimilar radio systems, and among secondary and primary radio systems. Cognitive radios are therefore most likely to be successful in short range, indoor communication scenarios, where mutual interference of other devices operating in the same frequency band is physically limited. Very high data rate communication without restrictive QoS constrains can be realized with cognitive radios. The regulation option of a licensed secondary radio system not requiring coordination support by primary radio systems is an attractive opportunity and is currently being discussed for the reuse of TV bands. It implies QoS guarantees due to a certain degree of exclusiveness of spectrum access and thus enables a reliable provisioning of billable services.

8

Concluding Remarks

This book was written in true excitement with the purpose of sharing our passion and expressing our enthusiasm for the fascinating research area of cognitive radio and dynamic spectrum access. We are confident that many research trends in this area, of which some have been discussed in this book, will change the wireless communication world and information society as we know it today.

We know that radio regulation has its own dynamics. Helping to change such dynamics is one of the main motivations for our book. Facilitating the decision making with cognitive radio is part of this change and will be a fruitful long term objective of wireless research. We are more than eager to follow such research activities in the years to come. In the meantime, enablers like the few we proposed in the preceding chapters may help the community to use spectrum more efficiently, without introducing more complexity. In this way, confidence in dynamic spectrum access and cognitive radio can grow on an incremental basis.

Future wireless broadband communications systems, covering wide geographic areas and nations, will be based on a tight combination of two ways of accessing spectrum. Firstly, exclusively accessed spectrum, and secondly, unlicensed open spectrum access, and both will coexist. The exclusively used spectrum will continue to allow operators to guarantee a minimum level of QoS in a conservative approach. The additional open spectrum, however, will enable the *ad-hoc* extension of network capacities, and coverage areas.

Intelligent spectrum sharing algorithms for mutual coordination and machine-understandable policies will improve the efficiency of spectrum usage. They will increase the capability of radio networks to support QoS by utilizing the open spectrum that is (vertically and horizontally) shared with competing networks.

All these approaches are part of our vision of dynamic spectrum access and cognitive radio. The great benefit lies in the flexibility it enables. Cognitive radio lies between the two ends of spectrum regulation: on the one hand, open spectrum, and on the other command-and-control for exclusive licensing. Cognitive radios can be modified to any level of freedom between these two extremes.

Liberating the access to spectrum will stimulate innovations and economic success. In distributed environments, adaptive cognitive radios guided by machine-understandable policy

---

descriptions will provide the necessarily required flexibility and intelligence for spectrum access.

The self-organization of cognitive radios to a cognitive radio network will further enhance efficiency and capacity, and will facilitate QoS support in wireless communication. A more flexible regulatory framework as discussed throughout the book will be needed to enable such a less-restricted spectrum usage.

Operator assistance may play an important role especially for in secondary spectrum usage with vertical spectrum sharing. Operators may assist in the identification of spectrum opportunities to protect incumbent radio systems. Operators may further control the cognitive radio networks to set up an optimal configuration dynamically.

The description of spectrum-sharing algorithms in a machine-understandable language is a challenging task that has to be supported with a policy language. To illustrate how such algorithms could be specified, the mapping of a spectrum-sharing algorithm to a policy description language is illustrated in this book. The distinction between spectrum opportunity and usage constraint facilitates a hierarchical structuring of an algorithm's policy description.

The type of wireless communication system that we call a true cognitive radio still requires a lot of research work to be performed. However, once established, such systems would have considerable implications for the regulation, standards, and technology applied today. Wireless communications systems will then coordinate the usage of radio spectrum themselves, without involving regulation at today's level. Self-organizing radio systems would then autonomously regulate in a technology-based approach: the machines would make the decisions, not humans. It can only be imagined how our economies could gain from such a flexible, technology-based approach. If successful, this approach would support new emerging wireless applications, while at the same time allowing existing incumbent radio services to continue operating without considerable quality-of-service degradation. Such an open approach would potentially significantly increase the usage of our radio spectrum. Cognitive radio networks will then complement the existing cellular networks operating in licensed frequency bands.

With cognitive radio, spectrum assignment and licensing will become more dynamic. Greater flexibility in responding to the emerging demands of the information society, as well as to market requirements, will be the important result.

Appendix A

'jemula802'

The software tool jemula802 is an open source JAVA class library developed and maintained by the authors of this book. The class libraries constitute a kernel for event-driven stochastic simulation referred to as 'jemula', plus the system model for IEEE 802 radio systems referred to as jemula802. The system model jemula802 requires methods and attributes of the kernel when performing stochastic evaluations. Hence, the kernel provides the core means for stochastic simulation.

The system model jemula802 models radio stations, traffic generators and a radio channel and is prepared to work with real-time systems as true implementation of 802 communication stacks. For this reason we usually refer to it as an 'emulator' instead of 'simulator.'

A.1 Software Architecture

The simulation core referred to as 'jemula' consists of three main packages called (1) kernel, (2) plot, and (3) statistics. The first package 'kernel' provides the classes for event handling and scheduling of events, including event, event handler, simulation time, and event scheduler. Note that simulations operate with their own simulated time, which is non-realtime and entirely independent of the time consumed during a simulation. For example, executing a system model for a simulated non-realtime of hundred milliseconds may require a computer to run for ten realtime seconds. The computation and memory power of the used machines and the number of events that would occur during the simulated time will determine the realtime duration of the simulation. If a larger number of events have to be scheduled and processed, the simulation will require longer times. The second and third packages, 'plot' and 'statistics,' provide the necessary classes to realize a stochastic evaluation and to illustrate the system status while the simulation is active.

The system model jemula802 contains additional packages. In these packages, the radio channel and all important protocol layers of the radio systems are specified. Here, according to the ISO/OSI reference model, radio channels in layer 0, the physical layer as layer 1, the medium access control layer as layer 2, plus the networking layer including traffic generators as layer 3 are all specified in independent packages. In addition, a station that comprises entities of layer 1, 2, and 3, is specified in an additional package of jemula802.

Cognitive Radio and Dynamic Spectrum Access L. Berlemann and S. Mangold
© 2009 John Wiley & Sons, Ltd

How many stations to be modelled, how the modelled protocol should be configured, and what type and amount of traffic to be emulated is described in the so-called XML scenario files that come with jemula802. In such scenario files, many parameters can be changed to accommodate specific needs of the model. For example, 802.11 stations could be modelled that support quality-of-service with basic protocol functionalities and multiple priorities.

The jemula802 code could be used as baseline for real-life implementations. It is implemented in modules that can be compiled into real-life systems. The event-handling of the non-realtime jemula core must be replaced by real events through interfaces. Further, since the original jemula802 code is using the non-realtime timers of the jemula core, such timers must be replaced with real timers. With such modifications, the protocol code of jemula802 is in principle usable for real systems. In addition, jemula802 can simply simulate a scenario by relying on the packages and interfaces with jemula core.

A.2 Availability and License

The code is freely available and thoroughly tested. It is periodically used during evaluation campaigns at teaching events and classes by the authors, for example when teaching mobile computing algorithms for quality-of-service in Wireless LAN 802.11. The tool is published 'as is' under the Berkeley Software Distribution (BSD) permissive free software license. The newer three-clause version of the BSD license, which does not require acknowledgement of original contributions when re-using the code, is applied. BSD is a Unix operating system derivative distributed by the Computer Systems Research Group at UC Berkeley. The BSD license, which was originally developed for this operating system, since then has been used in numerous other open source projects. It is here again used for the emulation tool. The license allows commercial use of jemula/jemula802, as well as jemula/jemula802 to be incorporated into commercial products. Redistribution and use in source and binary forms, with or without modification, are permitted provided that a source code will retain any existing copyright notice.

A.3 How to Obtain the Code and Get Started

The source code is hosted and maintained using Origo of ETH Zurich, Switzerland. Origo is an open source software development and collaboration platform allowing open- and closed-source projects. Both the two class libraries jemula and jemula802 can be obtained from there either through modern version control systems or as zipped download.

Using a state-of-the-art software development environment, the code can be compiled immediately. It relies on some standard JAVA classes, but all core functionalities such as stochastics and timing concepts are completely specified as part of the code itself.

In addition to the JAVA code, some XML scenario files are provided in the repository that can be used to specify basic scenarios. The graphical user interface provided in jemula802 plots the Wireless LAN 802.11 data and backoff timing exchanged between different stations.

Appendix B

'YouShi'

The software tool 'YouShi' has been developed by the authors for analyzing spectrum sharing of wireless networks and is based on the commercial software tool Matlab. Youshi is Chinese and stands for 'the game'. The results discussed in the Sections 5.5 and 6.2.6 of this book are produced with this tool in order to evaluate the spectrum sharing of cognitive radios. In effect, two approaches are realized in this tool: firstly, the application of solution concepts derived from game theory and Spectrum Load Smoothing (SLS). YouShi allows a stochastic evaluation of the observed QoS during a frame-based interaction. Secondly, in the application of game theory, the tool further offers an analytical calculation of the observed QoS in the spectrum sharing scenarios from Section 5.5.6. Additionally, YouShi enables the stochastic evaluation of the IEEE 802.11e EDCA in Berlemann *et al.* (2006d) through simulations. The capability to support QoS based on the EDCA and its backoff procedure is simulated.

B.1 Modeling QoS Requirements and Demands

Up to five cognitive radios are implemented in YouShi. An additional entity represents the EDCA of all spectrum sharing IEEE 802.11e-based cognitive radios as background traffic. The EDCA is modeled through an Ethernet traffic trace file (Bellcore, 2000) that is also used in the WARP2 simulator (Mangold, 2003). The trace file is logged at the Data Link Layer (DLL) and represents the offered traffic at the radio interface resulting from typical Internet applications. Due to it's lower priority access and the presence of multiple HCs allocating the medium, the EDCA is here overloaded.

The QoS requirements Θ, Δ and Ξ reflect the constraints to spectrum access of those applications that are supported by the cognitive radios. In the case of SLS these requirements are modeled as defined in Section 6.3.5.2, and for the game-theory based approach these QoS requirements are defined in a similar way in Section 5.5.5.1.

B.2 Resource Allocation and Collisions

The superframe of 802.11e is the basis for the interaction of the cognitive radios analyzed with YouShi. Within this book, a superframe duration of 100 ms is assumed independently of

Cognitive Radio and Dynamic Spectrum Access   L. Berlemann and S. Mangold
© 2009 John Wiley & Sons, Ltd

the spectrum sharing approach under consideration. In SLS, the 100 ms determine the duration of the periodic frame structure as illustrated in Figure B.1, while in the game model the duration of an SSG is defined by this superframe duration. Figure B.1 illustrates the resource allocation and collision resolution in a single frame from 100 ms to 200 ms in an HCCA spectrum sharing scenario as described in Section 6.3.6. One primary license holding radio system is represented by HC0. HC1 and HC2 apply SLS and distribute their allocations around HC0's allocations based on observation. A legacy EDCA entity accesses the medium with lower priority in case the medium is idle. The dotted lines mark the slots used for the SLS.

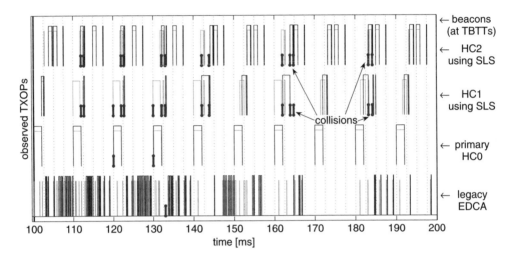

Figure B.1 Frame example from 100 ms to 200 ms of resource allocation modeled with YouShi2

Although both approaches developed in this book target the avoidance of collisions, cognitive radios may simultaneously attempt a transmission due to missing coordination. The collision of resource allocations is therefore modeled as follows. Each cognitive radio, representing a HC of 802.11e, has the same level of priority for allocating resources. In the case of the resource being idle after a transmission, the cognitive radios immediately attempt to access the resource before the lower priority EDCA entity tries spectrum allocation at a slightly later time. The allocation attempts of cognitive radios collide if multiple cognitive radios wait for the finalization of an ongoing transmission and simultaneously initiate their allocation attempt when the radio resource is idle again. In the case of a collision, the cognitive radios attempt transmission again in the order of their last successful resource allocation in the past. This resolves any collision between cognitive radios. Figure B.1 depicts the second step of the application of the SLS corresponding to Figure 6.18. The HCs are on the way to coordinate their allocations leading in later frames to a smoothed allocation distribution. The allocation attempts of the HCs collide here due to failing (inaccurate) coordination. Being based on the observation of past frames, the SLS using devices require some frames to distribute their allocations around the ones from the primary HC as collisions from the previous frame falsify the coordination in the current frame.

Additionally, the allocation attempt of the cognitive radios may also collide with attempts of the less prioritized EDCA entity, as also depicted in Figure B.1. In such a case, the cognitive radios obviously have the higher priority for accessing the radio resource after the collision. In the case of a collision thereafter, with an allocation attempt of another cognitive radio the mechanism from above is applied to resolve competition.

In YouShi spectrum sharing scenarios of completely overlapping WLANs operating at the same frequency, time and location are considered. An error-free channel model is assumed and side effects resulting from the hidden-station problem, link adaptation and power control are neglected.

B.3 Graphical User Interface

Figure B.2 depicts the graphical user interface of YouShi used for setting simulation and evaluation parameters.

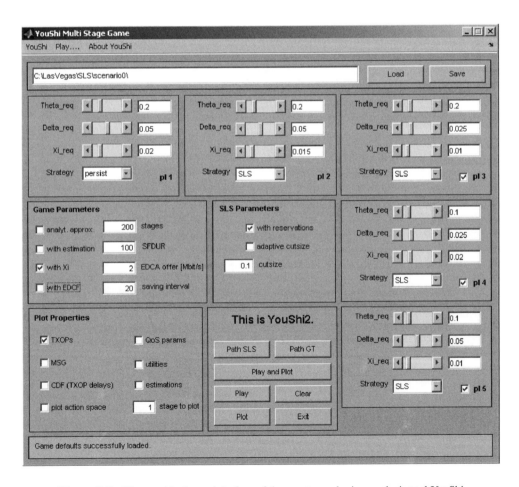

Figure B.2 The graphical user interface of the spectrum sharing analysis tool YouShi

References

Alonistioti, N., Glentis, A., Foukalas, F. and Kaloxylos, A. (2004). Reconfiguration management plane for the support of policy-based network reconfiguration. In *15th IEEE International Symposium on Personal, Indoor and Mobile Radio Communications*, PIMRC 2004, Barcelona, Spain, 5–8 September 2004.

Altman, E., Barman, D., Azouzi, R. E. and Jimenez, T. (2004). A Game Theoretic Approach for Delay Minimization in Slotted Aloha. In *IEEE International Conference on Communications*, ICC 2004, Paris, France, 20–24 June 2004.

Altman, E., Basar, T., Jimenez, T. and Shimkin, N. (2002). Competitive Routing in Networks with Polynomial Costs. *IEEE Transactions on Automatic Control*, **47** (1), 92–96.

Analysys (2004). Study on Conditions and Options in Introducing Secondary Trading of Radio Spectrum in the European Community. *Final Report for the European Commission*. Analysys Consulting Ltd (June 2005).

ARIB (2008). Available from: http://www.arib.or.jp/english/ (September 2008).

Axelrod, R. (1984). *The Evolution of Cooperation*. New York: Basic Books.

Baader, F., Calvanese, D., McGuinness, D., Nardi, D. and Patel-Schneider, P. (2003). *The Description Logic Handbook*. Cambridge University Press.

Bellcore (2000). Ethernet Tracefile pAug.TL. [online]. Bellcore (now Telcordia). Available from: ftp.telcordia.com/pub/world/wel/lantraffic (January 2003).

Berger, R. J. (2003). Open Spectrum: A Path to Ubiquitous Connectivity. ACM Queue, 1 (3). Available from: http://www.acmqueue.org/modules.php?name=Content&pa=showpage&pid=37&page=1.

Berlemann, L. (2002). Coexistence and Interworking in IEEE 802.11e in Competition Scenarios of Overlapping Wireless Networks. Thesis (Diploma), Chair of Communication Networks, RWTH Aachen University.

Berlemann, L., Mangold, S. and Walke, B. (2003a). Strategies, Behaviors, and Discounting in Radio Resource Sharing Games. In *10th Meeting of the Wireless World Research Forum*, WWRF10, New York USA, 27–28 October.

Berlemann, L., Siebert, M. and Walke, B. (2003b). Software Defined Protocols Based on Generic Protocol Functions for Wired and Wireless Networks. In *Software Defined Radio Technical Conference*, SDR'03, Orlando USA, 17–19 November.

Berlemann, L. (2003). Game Theoretical Modeling and Evaluation of the Radio Resource Competition of IEEE 802.11e Wireless Networks. Thesis (Diploma), Institute for Microeconomics, RWTH Aachen University.

Berlemann, L., Walke, B. and Mangold, S. (2004a). Behavior Based Strategies in Radio Resource Sharing Games. In *15th IEEE International Symposium on Personal, Indoor and Mobile Radio Communications*, PIMRC 2004, Barcelona, Spain, 5–8 September, pp. 840–846.

Berlemann, L., Hiertz, G. R., Walke, B. and Mangold, S. (2004b). Cooperation in Radio Resource Sharing Games of Adaptive Strategies. In *IEEE 60th Vehicular Technology Conference*, VTC2004–Fall, Los Angeles CA, USA, 26–29 September, pp. 3004–3009.

Berlemann, L. and Mangold, S. (2005). Policy-based Spectrum Navigation of Cognitive Radios in Open Spectrum. In *14th Meeting of the Wireless World Research Forum*, WWRF14, San Diego CA, USA, 7–8 July.

Berlemann, L., Hiertz, G. R., Walke, B. and Mangold, S. (2005a). Strategies for Distributed QoS Support in Radio Spectrum Sharing. In *IEEE International Conference on Communications*, ICC 2005, Seoul, Korea, 16–20 May, pp. 3271–3277.

Berlemann, L., Hiertz, G. R. and Walke, B. (2005b). Reservation-based Spectrum Load Smoothing as Cognitive Medium Access for Spectrum Sharing Wireless Networks. In *European Wireless 2005* , EW'05, Nicosia, Cyprus, 10–13 April, pp. 547–553.

Berlemann, L., Mangold, S., Hiertz, G. R. and Walke, B. (2005c). Radio Resource Sharing Games: Enabling QoS Support in Unlicensed Bands. *IEEE Network*, Special Issue on 'Wireless Local Area Networking: QoS Provision and Resource Management', **19** (4), 59–65. Available from: http://www.comsoc.org/ni/Public/2005/jul/, (August 2005).

Berlemann, L., Mangold, S., Hiertz, G. R. and Walke, B. H. (2005d). Spectrum Load Smoothing for Cognitive Medium Access in Open Spectrum. In *16th IEEE Conference on Personal, Indoor and Mobile Radio Communications*, PIMRC 2005, Berlin, Germany, 11–14 September.

Berlemann, L. and Walke, B. (2005). Spectrum Load Smoothing for Optimized Spectrum Utilization – Rationale and Algorithm. In *IEEE Wireless Communication and Networking Conference*, WCNC 2005, New Orleans LA, USA, 13–17 March 2005.

Berlemann, L., Mangold, S. and Walke, B. H. (2005e). Policy-based Reasoning for Spectrum Sharing in Cognitive Radio Networks. In *1st IEEE International Symposium on New Frontiers in Dynamic Spectrum Access Networks*, DySPAN2005, Baltimore MD, USA, 8–11 November, pp. 1–10.

Berlemann, L. (2006). Distributed Quality-of-Service Support in Cognitive Radio Networks. Thesis (PhD), Chair of Communication Networks, RWTH Aachen University, Aachen Germany: Wissenschaftsverlag Mainz.

Berlemann, L., Mangold, S., Jarosch, A. and Walke, B. (2006a). Cognitive Radio: Business Impacts of Dynamic Spectrum Access and Spectrum Sharing, in *VDE Congress 2006*. Aachen, Germany.

Berlemann, L., Hoymann, C., Hiertz, G. R. and Mangold, S. (2006b). Coexistence and Interworking of IEEE 802.16 and IEEE 802.11(e). In *IEEE 63rd Vehicular Technology Conference*, VTC2006, Spring, Melbourne, Australia, 7–10 May.

Berlemann, L., Mangold, S., Hiertz, G. R. and Walke, B. H. (2006c). Spectrum Load Smoothing: Distributed Quality-of-Service Support for Cognitive Radios in Open Spectrum. *European Transactions on Telecommunications*, **17** (3), 395–406.

Berlemann, L., Mangold, S., Hiertz, G. R. and Walke, B. H. (2006d). Delay Performance of the Enhanced Distributed Channel Access of IEEE 802.11e. In *15th IST Mobile & Wireless Communications Summit*, Mykonos, Greece, 4–8 June.

Biron, P. V., Permanente, K. and Malhotra, A. (2004). *XML Schema Part 2: Datatypes*, Second Edition, W3C Recommendation [online]. World Wide Web Consortium. Available from: http://www.w3.org/TR/2004/REC–xmlschema–2–20041028/ (June 2005).

Bourse, D., El-Khazen, K., Delautre, A., Wiebke, T., Dillinger, M., Brakensiek, J., Moessner, K., Vivier, G. and Alonistioti, N. (2004). European Research in End-to-End Reconfigurability. In *13th IST Mobile & Wireless Communications Summit* 2004, Lyon, France 27–30 June.

Bourse, D., El-Khazen, K., Berthet, N., Wiebke, T., Dillinger, M., Brakensiek, J., Moessner, K., Vivier, G. and Alonistioti, N. (2005). End-to-End Reconfigurability: Towards the Seamless Experience. In *14th IST Mobile & Wireless Communications Summit 2005*, Dresden Germany, 19–23 June.

Buddhikot, M. M., Kolodzy, P., Miller, S., Ryan, K. and Evans, J. (2005). DIMSUMNet: New Directions in Wireless Networking Using Coordinated Dynamic Spectrum Access (position paper). In *IEEE International Symposium on a World of Wireless, Mobile and Multimedia Networks*, WoWMoM 2005, Taormina, Italy, 13–16 June.

Bundesnetzagentur (2005). Strategic Aspects of the Spectrum Regulation of the RegTP [online]. Available from: http://www.bundesnetzagentur.de/media/archive/2155.pdf].

Cabric, D., Mishra, S. M., Willkomm, D., Brodersen, R. and Wolisz, A. (2005). A Cognitive Radio Approach for Usage of Virtual Unlicensed Spectrum. In *14th IST Mobile & Wireless Communications Summit*, Dresden, Germany, 19–23 June.

Cagalj, M., Ganeriwal, S., Aad, I. and Hubaux, J.-P. (2005). On Selfish Behavior in CSMA/CA Networks. In *IEEE INFOCOM 2005*, Miami, USA, 13–17 March.

Clark, D. D., Partridge, C., Ramming, J. C. and Wroclawski, J. T. (2003). A Knowledge Plane for the Internet. In *ACM Special Interest Group on Data Communications Conference*, ACM SIGCOMM 2003, Karlsruhe, Germany, August.

Coase, R. H. (1959). The Federal Communications Commission. *Journal of Law and Economics*, **2**, 1–40.

Coase, R. H. (1960). The Problem of Social Cost. *Journal of Law and Economics*, **3**, 1–44.

Courcoubetis, C. and Webber, R. (2003). *Pricing Communication Networks – Economics, Technology and Modelling*. Chichester, UK: John Wiley & Sons, Ltd.

Cover, T. M. and Thomas, J. A. (1991). *Elements of Information Theory*. New York, USA: John Wiley & Sons, Inc.

DARPA (2003). The XG Architectural Framework. Request for Comments. Version 1.0. Prepared by BBN Technologies, Cambridge MA, USA: DARPA XG Working Group. Available from: http://www.ir.bbn.com/projects/xmac/vision.html (May 2005).

DARPA (2004a). XG Policy Language Framework. Request for Comments. Version 1.0. Prepared by BBN Technologies, Cambridge MA, USA: DARPA XG Working Group. Available from: http://www.ir.bbn.com/projects/xmac/pollang.html (May 2005).

DARPA (2004b). The XG Vision. Request for Comments. Version 2.0. Prepared by BBN Technologies, Cambridge MA, USA: DARPA XG Working Group. Available from: http://www.ir.bbn.com/projects/xmac/vision.html (May 2005).

DARPA (2008). Available from: http://www.darpa.mil/sto/smallunitops/xg.html (September 2008).

Das, S. K., Lin, H. and Chatterjee, M. (2004). An Econometric Model for Resource Management in Competitive Wireless Data Networks. *IEEE Network*, **18** (6), 20–26.

Debreu, D. (1959). *Theory of Value. An axiomatic Analysis of Economic Equilibrium*. New Haven, USA and London, UK: Yale University Press.

Demestichas, P., Koutsouris, N., KoundourakiS, G., Tsagkaris, K., Oikonomou, A., Stavroulaki, V., Papadopoulou, L., Theologou, M. E., vivier, G. and EL Khazen, K. (2003). Management of Networks and Services in a Composite Radio Context. *IEEE Wireless Communications*, **10** (4), 44–51.

Demestichas, P., Vivier, G., El-Khazen, K. and Theologou, M. (2004). Evolution in Wireless Systems Management Concepts: From Composite Radio Environments to Reconfigurability. *IEEE Communications Magazine*, **42** (5), 90–98.

Dillinger, M., Madani, K. and Alonistioti, N. (2003). Software Defined Radios: Architectures, Systems and Functions. Chichester, UK: John Wiley & Sons, Ltd.

Dutta-Roy, A. (2000). The Cost of Quality in Internet–style Networks. *IEEE Spectrum*, **37** (9), 57–62.

E3 (2008). Available from: https://www.ict–e3.eu/ (September 2008).

EC (2007). *Rapid Access to Spectrum for Wireless Electronic Communication Services through more Flexibility*. Brussels: European Commission.

ECC (2004). ECC Decision of 12 November 2004 on the harmonised use of the 5 GHz frequency bands for the implementation of Wireless Access Systems including Radio Local Area Networks (WAS/RLANs). ECC/DEC/(04)08. Electronic Communications Committee.

Elenius, D., Denker, G., Stehr, M.-O., Senanayake, R., Talcott, C. and Wilkins, D. (2007). CoRaL – Policy Language and Reasoning Techniques for Spectrum Policies. In *IEEE International Workshop on Policies for Distributed Systems and Networks*, June 13–15, pp. 261–265.

ERC (2001). ERC Decision of 12 March 2001 on harmonised frequencies, technical characteristics and exemption from individual licensing of Short Range Devices used for Radio Local Area Networks (RLANs) operating in the frequency band 2400 – 2483.5 MHz. ERC/DEC 01–07. European Radiocommunications Committee.

ERC (2005). ERC Recommendation 70–03 Relating to the Use of Short Range Devices (SRD). ERC/REC 70–03. European Radiocommunications Committee.

EU (2002a). Decision No 676/2002/EC on a regulatory framework for radio spectrum policy in the European Community (Radio Spectrum Decision). The European Parliament and the Council of the European Union. Available from: http://www.euractiv.com/Article?tcmuri=tcm:29–117547–16&type=LinksDossier.

EU (2002b). Directive 2002/21/EC on a common regulatory framework for electronic communications networks and services (Framework Directive). The European Parliament and the Council of the European Union. Available from: http://www.euractiv.com/Article?tcmuri=tcm:29–117547–16&type=LinksDossier.

Evans, J., Minden, G. and Knightly, E. (2006). Cognitive Radio Networks in GENI. GENI Design Document 06–20. GENI Wireless Working Group.

FCC (2002). Report of the Spectrum Rights and Responsibilities Working Group. ET Docket No. 02–135. Federal Communications Commission Spectrum Policy Task Force.

FCC (2003a). Report and Order (FCC 03–287): Revision of Parts 2 and 15 of the Commission's Rules to Permit Unlicensed National Information Infrastructure (U–NII) devices in the 5 GHz band. ET Docket No. 03–122. Federal Communications Commission.

FCC (2003b). Notice of Inquiry and Notice of Proposed Rulemaking (FCC 03–289): Establishment of an Interfer-
    ence Temperature Metric to Quantify and Manage Interference and to Expand Available Unlicensed Operation
    in Certain Fixed, Mobile and Satellite Frequency Bands. ET Docket No. 03–237. Federal Communications
    Commission.

FCC (2003c). Notice for Proposed Rulemaking (FCC 03–322): Facilitating Opportunities for Flexible, Effi-
    cient, and Reliable Spectrum Use Employing Cognitive Radio Technologies. ET Docket No. 03–108. Federal
    Communications Commission.

FCC (2004a). Second Report and Order (FCC 04–167): Promoting Efficient Use of Spectrum Through Elimina-
    tion of Barriers to the Development of Secondary Markets. WT Docket No. 00–230. Federal Communications
    Commission.

FCC (2004b). Notice of Proposed Rulemaking (FCC 04–100): Unlicensed Operation in the Band 3650–3700 MHz.
    ET Docket No. 04–151. Federal Communications Commission.

FCC (2004c). Notice of Proposed Rulemaking (FCC 04–113): Unlicensed Operation in the TV Broadcast Bands. ET
    Docket No. 04–186. Federal Communications Commission.

FCC (2005a). Report and Order and Memorandum Opinion and Order (FCC 05–56): Wireless Operations in the
    3650–3700 MHz Band. ET Docket No. 04–151. Federal Communications Commission.

FCC (2005b). Report and Order (FCC 05–57): Facilitating Opportunities for Flexible, Efficient, and Reliable
    Spectrum Use Employing Cognitive Radio Technologies. ET Docket No. 03–108. Federal Communications
    Commission.

FCC (2005c). Part 15 – Radio Frequency Devices – of the Commissions Rules to Permit Unlicensed National Infor-
    mation Infrastructure (U NII) devices in the 5 GHz band. Federal Communications Commission. Available from:
    http://www.fcc.gov/oet/info/rules/part15/part15_4_05_05.pdf.

FCC (2007). Memorandum Opinion and Order (FCC 07–99). Federal Communications Commission.

Félegyházi, M., Hubaux, J.-P. and Buttyán, L. (2005). Nash Equilibria of Packet Forwarding Strategies in Wireless
    Ad Hoc Networks. To appear in IEEE Transactions on Mobile Computing (TMC).

Friedman, J. W., (1971). A Non–cooperative Equilibrium for Supergames. Review of Economic Studies, 38.

Fudenberg, D. and Tirole, J., (1998). Game Theory. Cambridge, Massachusetts, USA and London, UK: MIT Press.

Fudenberg, D. and Levine, D. K. (1998). The Theory of Learning in Games. Cambridge, Massachusetts, USA and
    London, UK: MIT Press.

Gallager, R. G. (1968). Information Theory and Reliable Communication. New York: John Wiley & Sons, Inc.

GAMETHEORY.NET, (2005). Available from: http://www.gametheory.net/ (October).

GENI, (2008). Available from: http://www.geni.net/ (September 2008).

Grandblaise, D., Kloeck, C., Moessner, K., Mohyeldin, E., Pereirasamy, M. K., Luo, J. and Martoyo, I. (2005).
    Towards a Cognitive Radio based Distributed Spectrum Management. In 14th IST Mobile & Wireless Communi-
    cations Summit, Dresden, Germany, 19–23 June.

Halldorsson, M. M., Halpern, J. Y., Li, L. and Mirrokni, V. (2004). On Spectrum Sharing Games. In 23rd Annual
    ACM Symposium on Principles of Distributed Computing, PODC 2004, St. John's, Newfoundland, Canada, 25–28
    July, pp. 107–114.

Hardin, G., (1968). The Tragedy of the Commons. Science, 162 (3859), 1243–1248. Available from: http://
    www.garretthardinsociety.org/articles/art_tragedy_of_the_commons.html.

Haykin, S. (2005). Cognitive Radio: Brain–Empowered Wireless Communications. IEEE Journal on Selected Areas
    in Communications, 23 (2), 201–220.

Hazlett, T. W. (1998). Spectrum Flash Dance: Eli Noam's Proposal for 'Open Access' to Radio Waves. Journal of
    Law and Economics, 44, 805–821.

Hazlett, T. W. (2005). Spectrum Tragedy. Yale Journal on Regulation, 22.

Hiertz, G. R., Habetha, J., May, P., Weiss, E., Bagul, R. and Mangold, S. (2003). A Decentralized Reservation
    Scheme for IEEE 802.11 Ad Hoc Networks. In 14th IEEE Conference on Personal, Indoor and Mobile Radio
    Communications, PIMRC 2003, Beijing, China, 7–10 September.

Hiertz, G. R., Zang, Y., Habetha, J. and Sirin, H. (2005a). Multiband OFDM Alliance – The Next Generation of
    Wireless Personal Area Networks. In 2005 IEEE Sarnoff Symposium, Princeton NJ, USA, April.

Hiertz, G. R., Zang, Y., Habetha, J. and Sirin, H. (2005b). IEEE 802.15.3a Wireless Personal Area Networks –
    The MBOA Approach. In 11th European Wireless Conference 2005, EW'05, Nicosia, Cyprus, 10–13 April,
    pp. 204–210.

Hiertz, G. R., Zang, Y., Stibor, L., Max, S., Reumerman, H.-J., Sanchez, D. and Habetha, J. (2005c). Mesh Networks
    Alliance (MNA) – Proposal – IEEE 802.11s – MAC Sublayer Functional Description – IEEE 802.11s – Mesh

WLAN Security (IEEE 802.11 Proposal Submission). In *802.11 WLAN Working Group Session*, San Francisco CA, USA, July 2005c.

Hiertz, G. R., Denteneer, D., Max, S., Taori, R., Cardona, J., Berlemann, L. and Walke, B. (2008a). IEEE 802.11s – The WLAN Mesh Standard. *IEEE Wireless Communications Magazine*.

Hiertz, G. R., Zang, Y., Max, S., Junge, T., Weiss, E., Wolz, B., Denteneer, D., Berlemann, L. and Mangold, S. (2008b). IEEE 802.11s: WLAN Mesh Standardization and High Performance Extensions. *IEEE Network*, **22** (3), 12–19.

Horne, W. D. (2003). Adaptive Spectrum Access: Using the Full Spectrum Space. In *31st Research Conference on Communication, Information and Internet Policy*, TPRC'03, Arlington VA, USA, 19–21 September.

Hulpert, A. P. (2005). Spectrum Sharing Through Beacons. In *16th IEEE Conference on Personal, Indoor and Mobile Radio Communications*, PIMRC 2005, Berlin, Germany, 11–14 September.

Hunold, D., Barreto, A. N., Fettweis, G. P. and Mecking, M. (2000). Concept for Universal Access and Connectivity in Mobile Radionetworks. In *11th IEEE International Symposium on Personal, Indoor and Mobile Radio Communications*, PIMRC 2000, London, UK, 18–21 September, pp. 847–851.

IEEE (1998). IEEE Standard for Information Technology Telecommunications and Information Exchange between Systems Local and Metropolitan Area Networks Common Specifications Part 3: Media Access Control (MAC). IEEE 802.1D. New York, USA: Institute of Electrical and Electronics Engineers, Inc.

IEEE (1999). Supplement to IEEE Standard for Information Technology – Telecommunications and Information Exchange Between Systems – Local and Metropolitan Area Networks – Specific Requirements – Part 11: Wireless LAN Medium Access Control (MAC) and Physical Layer (PHY) specifications: High-speed Physical Layer in the 5 GHZ Band. IEEE Std 802.11a–1999. New York, USA: Institute of Electrical and Electronics Engineers, Inc.

IEEE (2002). IEEE Standard for Local and Metropolitan Area Networks, Part 16: Air Interface for Fixed Broadband Wireless Access Systems. IEEE Std 802.16–2001. New York, USA: Institute of Electrical and Electronics Engineers, Inc.

IEEE (2004a). Draft Amendment to STANDARD FOR Telecommunications and Information Exchange Between Systems – LAN/MAN Specific Requirements – Part 11: Wireless Medium Access Control (MAC) and Physical Layer (PHY) Specifications: Medium Access Control (MAC) Quality of Service (QoS). IEEE 802.11e/ D8.0. New York, USA: Institute of Electrical and Electronics Engineers, Inc.

IEEE (2004b). Recommended Practice for Local and metropolitan area networks – Coexistence of Fixed Broadband Wireless Access Systems. IEEE Std 802.16.2–2004. New York, USA: Institute of Electrical and Electronics Engineers, Inc.

IEEE (2005a). Amendment to IEEE Standard for Local and Metropolitan Area Networks – Part 16: Air Interface for Fixed Broadband Wireless Access Systems – Improved Coexistence Mechanisms for License–Exempt Operation. Working document for IEEE 802.16h. New York, USA: Institute of Electrical and Electronics Engineers, Inc. Available from: http://ieee802.org/16/le/docs/80216h–05_013.pdf.

IEEE (2005b). IEEE Standard for Information Technology – Telecommunications and Information Exchange Between Systems – LAN/MAN Specific Requirements – Part 11: Wireless LAN Medium Access Control (MAC) and Physical Layer (PHY) Specifications: Amendment: Medium Access Control (MAC) Quality of Service Enhancements. IEEE 802.11e/ D13.0. New York, USA: Institute of Electrical and Electronics Engineers, Inc.

IEEE (2005c). Draft Amendment to STANDARD FOR Information Technology – Telecommunications and Information Exchange Between Systems – LAN/MAN Specific Requirements – Part 11: Wireless Medium Access Control (MAC) and Physical Layer (PHY) Specifications: Amendment 7: Radio Resource Measurement. IEEE 802.11k/D2.0. New York, USA: Institute of Electrical and Electronics Engineers, Inc.

IEEE (2006). Draft PHY/MAC Specification for IEEE 802.22. IEEE 802.22–06/0069r1. New York, USA: Institute of Electrical and Electronics Engineers, Inc.

IEEE (2007). IEEE Standard for Information Technology – Telecommunications and Information Exchange Between Systems – Local and Metropolitan Area Networks – Specific Requirements Part 11: Wireless LAN Medium Access Control (MAC) and Physical Layer (PHY) Specifications. IEEE Std 802.11–2007. New York, USA: Institute of Electrical and Electronics Engineers, Inc.

IEEE (2008a). Draft Amendment to STANDARD FOR Information Technology – Telecommunications and Information Exchange Between Systems – LAN/MAN Specific Requirements – Part 11: Wireless Medium Access Control (MAC) and Physical Layer (PHY) Specifications: Amendment 3: 3650–3700 MHz Operation in USA. P802.11y/D11.0. New York, USA: Institute of Electrical and Electronics Engineers, Inc.

IEEE (2008b). Draft Amendment to Standard for Information Technology – Telecommunications and Information Exchange Between Systems – LAN/MAN Specific Requirements – Part 11: Wireless Medium Access Control

(MAC) and Physical Layer (PHY) Specifications – Amendment 5: Enhancements for Higher Throughput. IEEE P802.11n/D6.0. New York, USA: Institute of Electrical and Electronics Engineers, Inc.

IEEE (2008c). Draft Amendment to Standard for Information Technology – Telecommunications and Information Exchange Between Systems – LAN/MAN Specific Requirements – Part 11: Wireless Medium Access Control (MAC) and Physical Layer (PHY) Specifications: Amendment: Mesh Networking. IEEE P802.11s/D2.0. New York, USA: Institute of Electrical and Electronics Engineers, Inc.

Kasturia, S., Aslanis, J. and Cioffi, J. M. (1990). Vector Coding for Partial-Response Channels. *IEEE Transactions on Information Theory*, **36** (4), 741–762.

Kloeck, C., Jaekel, H. and Jondral, F. K. (2005). Auctions Sequence as a New Spectrum Allocation Mechanism. In *14th IST Mobile & Wireless Communications Summit*, Dresden, Germany, 19–23 June.

Kreps, D. (1990). *A Course in Microeconomic Theory*. Princeton NJ, USA: Princeton University Press.

Kruys, J. (2003). Co-existence of Dissimilar Wireless Systems [online]. Available from: http://www.wi–fi.org/opensection/pdf/co–existence_dissimilar_systems.pdf (July 2005).

Lansford, J. (2004). UWB Coexistence and Cognitive Radio In *International Workshop on Ultra Wideband Systems*, 2004, Kyoto, Japan, 18–21 May, pp. 35–39.

Lucky, R. W. (2001). The Precious Radio Spectrum. *IEEE Spectrum*, **38** (9), 90.

Maheswaran, R. T. and Basar, T. (2003). Nash Equilibrium and Decentralized Negotiation in Auctioning Divisible Resources. *Group Decision and Negotiation*, **12** (5), 361–395.

Mähönen, P. (2004). Cognitive Trends in Making: Future of Networks. In *15th IEEE International Symposium on Personal, Indoor and Mobile Radio Communications*, PIMRC 2004, Barcelona, Spain, 5–8 September, pp. 1449–1454.

Mähönen, P., Riihijarvi, J., Petrova, M. and Shelby, Z. (2004). Hop-by-hop toward Future Mobile Broadband IP. *IEEE Communications Magazine*, **42** (3), 138–146.

Mangold, S., Habetha, J., Choi, S. and Ngo, C. (2001a). Co-existence and Interworking of IEEE 802.11a and ETSI BRAN HiperLAN/2 in MultiHop Scenarios. In *IEEE 3rd Workshop in Wireless Local Area Networks*, Boston, USA, 27–28 September.

Mangold, S., Choi, S., Budde, W. and Ngo, C. (2001b). 802.11a/e and H2 Interworking Using HCF. IEEE Working Document 802.11–01/414r0. Available from: http://www.ieee802.org/11.

Mangold, S., Berlemann, L. and Hiertz, G. R. (2002). QoS Support as Utility for Coexisting Wireless LANs. In *International Workshop on IP Based Cellular Networks*, IPCN'02, Paris, France, 23–26 April.

Mangold, S. (2003). Analysis of IEEE 802.11e and Application of Game Models for Support of Quality-of-Service in Coexisting Wireless Networks Thesis (PhD), Chair of Communication Networks, RWTH Aachen University, Aachen Germany: Wissenschaftsverlag Mainz.

Mangold, S., Berlemann, L. and Walke, B. (2003b). Radio Resource Sharing Model for Coexisting IEEE 802.11e Wireless LANs. In *International Conference on Communication Technology Proceedings*, ICCT 2003, Beijing, China, 9–11 April 2003b, 1322–1327.

Mangold, S. and Challapali, K. (2003). Coexistence of Wireless Networks in Unlicensed Frequency Bands. In *9th Meeting of the Wireless World Research Forum*, WWRF9, Zurich, Switzerland, July.

Mangold, S., Choi, S., Hiertz, G. R., Klein, O. and Walke, B. (2003a). Analysis of IEEE 802.11e for QoS Support in Wireless LANs. *IEEE Wireless Communications*, **10** (6), 40–50.

Mangold, S., Berlemann, L. and Walke, B. (2003c). Equilibrium Analysis of Coexisting IEEE 802.11e Wireless LANs. In *14th IEEE Conference on Personal, Indoor and Mobile Radio Communications*, PIMRC 2003, Beijing, China, 7–10 September, pp. 321–325.

Mangold, S., Zhong, Z., Hiertz, G. R. and Walke, B. (2004a). IEEE 802.11e/802.11k Wireless LAN: Spectrum Awareness for Distributed Resource Sharing. *Wireless Communications and Mobile Computing*, **4** (8), 881–902.

Mangold, S., Zhong, Z., Challapali, K. and Chou, C. T. (2004b). Spectrum Agile Radio: Radio Resource Measurements for Opportunistic Spectrum Usage. In *47th annual IEEE Global Telecommunications Conference*, Globecom 2004, Dallas TX, USA, 29 November – 3 December 2004b.

Mangold, S., Shankar, S. and Berlemann, L. (2005a). Spectrum Agile Radio: A Society of Machines with Value-Orientation (invited paper). In *European Wireless Conference 2005*, EW'05, Nicosia, Cyprus, 10–13 April, pp. 539–546.

Mangold, S. and Berlemann, L. (2005). IEEE 802.11k: Improving Confidence in Radio Resource Measurements. In *16th IEEE Conference on Personal, Indoor and Mobile Radio Communications*, PIMRC 2005, Berlin, Germany, 11–14 September.

Mangold, S. and Habetha, J. (2005). IEEE 802.11 Contention-Based Medium Access for Multiple Channels. In *11th European Wireless Conference 2005*, Nicosia, Cyprus, April, pp. 522–526.

Mangold, S., Jarosch, A. and Monney, C. (2005b). Cognitive Radio – Trends and Research Challenges. *Swisscom Comtec Magazine* 03/2005, 6–9.

Mangold, S., Berlemann, L. and Nandagopalan, S. (2006a). Spectrum Sharing with Value-Orientation for Cognitive Radio. *European Transactions on Telecommunications*, **17** (3), 383–394.

Mangold, S., Jarosch, A. and Monney, C. (2006b). Operator Assisted Cognitive Radio and Dynamic Spectrum Assignment with Dual Beacons – Detailed Evaluation. In *First International Workshop on Wireless Personal and Local Area Networks*, New Delhi, India, 8 January.

Mangold, S., Jarosch, A. and Monney, C. (2006c). Operator Assisted Dynamic Spectrum Assignment with Dual Beacons. In *IEEE International Zurich Seminar on Communications* IZS, Zurich, Switzerland, 22–24 February.

Mas-Colell, A., Whinston, M. D. and Green, J. (1997). *Microeconomic Theory*. New York, USA: Oxford University Press.

Max, S., Hiertz, G. R., Weiss, E., Denteneer, D. and Walke, B. (2006). Spectrum Sharing in IEEE 802.11s Wireless Mesh Networks. *Computer Networks*, **51** (9), 2353–2367.

May, R. J., Speta, J. B., Dixon, K. D., Gattuso, J. L., Gifford, R. L., Shelanski, H. A., Sicker, D. C. and Weisman, D. (2005). *Digital Age Communications Act – A Proposal for a New Regulatory Framework* – Release 1.0. Washington DC, USA: Progress & Freedom Foundation.

McGuinness, D. L. and Harmelen, F. V. (2003). OWL Web Ontology Language Overview, W3C Candidate Recommendation [online]. World Wide Web Consortium. Available from: http://www.w3.org/TR/2003/CR–owl–features–20030818/ (June 2005).

Mehta, M., Drew, N., Vardoulias, G., Greco, N. and Niedermeier, C. (2001). Reconfigurable Terminals: An Overview of Architectural Solutions. *IEEE Communications Magazine*, **39** (8), 82–89.

MIC (2003). *Radio Policy Vision: Report from Telecommunications Council*. Ministry of Internal Affairs and Communications.

Mitola, J., III (1995). The Software Radio Architecture. *IEEE Communications Magazine*, **33** (5), 26–38.

Mitola, J., III and Maguire, G. Q., JR. (1999). Cognitive Radio: Making Software Radios More Personal. *IEEE Personal Communications Magazine*, **6** (4), 13–18.

Mitola, J., III (2000). Cognitive Radio: An Integrated Agent Architecture for Software Defined Radio. Thesis (PhD), Dept. of Teleinformatics, Royal Institute of Technology (KTH), Stockholm, Sweden.

Mohr, W. (2005). The WINNER (Wireless World Initiative New Radio) Project – Development of a Radio Interface for Systems beyond 3G In *16th IEEE Conference on Personal, Indoor and Mobile Radio Communications*, PIMRC 2005, Berlin, Germany, 11–14 September.

Nash, J. F. (1950). Non-cooperative Games. Thesis (PhD), Princton University.

Neel, J., Buehrer, R. M., Reed, B. H. and Gilles, R. P. (2002a). Game Theoretic Analysis of a Network of Cognitive Radios. In *The 2002 45th Midwest Symposium on Circuits and Systems*, MWSCAS–2002, 4–7 August, pp. 409–412.

Neel, J., Reed, J. and Gilles, R. (2002b). The Role of Game Theory in the Analysis of Software Radio Networks. In *SDR Forum Technical Conference and Product Exhibition*, SDR'02, San Diego CA, USA, November.

Neel, J. O., Reed, J. H. and Gilles, R. P. (2004). Convergence of Cognitive Radio Networks In *IEEE Wireless Communications and Networking Conference*, WCNC 2004 New Orleans LA, USA, 21–25 March, pp. 2250–2255.

NGMN (2007). *Use Cases related to Self Organising Network. Overall Description*. Version 2.02. Next Generation Mobile Networks Alliance.

Nie, N. and Comaniciu, C. (2005). Adaptive Channel Allocation Spectrum Etiquette for Cognitive Radio Networks. In *1st IEEE International Symposium on New Frontiers in Dynamic Spectrum Access Networks*, DySPAN, Baltimore MD, USA, 8–11 November.

Noam, E. (1997). Beyond Spectrum Auctions: Taking the Next Step to Open Spectrum Access. *Telecommunications Policy*, **21** (5), 461–475.

Noam, E. (1998). Spectrum Auctions: Yesterday's Heresy, Today's Orthodoxy, Tomorrow's Anachronism. Taking the Next Step to Open Spectrum Access. *Journal of Law and Economics*, **44**, 765–790.

OFCOM, (2004). Spectrum Framework Review. UK, London: Office of Communications.

OFCOM (2008). *Digital Dividend Review: 550–630 MHz and 790–854 MHz*. UK, London: Office of Communications.

ORACLE (2008). Available from: http://www.ist–oracle.org/ (September 2008).

Osborne, M. J. and Rubinstein, A. (1994). *A Course in Game Theory*. Cambridge, Massachusetts, USA and London, UK: The MIT Press.

Peha, J. M. (1998). *Spectrum Management Policy Options*. IEEE Communication Surveys.

Peha, J. M. (2000). Wireless Communications and Coexistence for Smart Environments. *IEEE Personal Communications*, 66–68.

Peha, J. M. (2005). Approaches to Spectrum Sharing. *IEEE Communications Magazine*, 10–12.

Pereirasamy, M. K., Luo, J., Dillinger, M. and Hartmann, C. (2004). An Approach for Inter-Operator Spectrum Sharing for 3G Systems and Beyond. In *15th IEEE International Symposium on Personal, Indoor and Mobile Radio Communications*, PIMRC 2004, Barcelona, Spain, 5–8 September.

Pereirasamy, M. K., Luo, J., Dillinger, M. and Hartmann, C. (2005). Dynamic Inter-Operator Spectrum Sharing for UMTS FDD with Displaced Cellular Networks. In *IEEE Wireless Communication and Networking Conference*, WCNC 2005, New Orleans, LA, USA, 13–17 March.

Pfletschinger, S., Hooli, K., Huschke, J., Mohr, W. and Ojanen, P. (2005). Possible Impact of Emissions from UWB Devices on Systems beyond 3G. In *14th Meeting of the Wireless World Research Forum*, WWRF14, San Diego CA, USA, 7–8 July.

Popescu, D. C. (2002). Interference Avoidance for Wireless Systems. Thesis (PhD), Department of Electrical and Computer Engineering, Rutgers University.

Popescu, D. C. and Rose, C. (2004). *Interference Avoidance Methods for Wireless Systems*. New York, USA: Kluwer Academic Publishers.

Raychaudhuri, D. and Jing, X. (2003). A Spectrum Etiquette Protocol for Efficient Coordination of Radio Devices in Unlicensed Bands. In *14th IEEE Conference on Personal, Indoor and Mobile Radio Communications*, PIMRC 2003, Beijing, China, 7–10 September.

ROCKET (2008). Available from: http://www.ict–rocket.eu/ (September 2008).

Rodriguez, V., Moessner, K. and Tafazolli, R. (2005). Auction Driven Dynamic Spectrum Allocation: Optimal Bidding, Pricing and Service Priorities for Multi-rate, Multi-Class CDMA. In *16th IEEE Conference on Personal, Indoor and Mobile Radio Communications*, PIMRC 2005, Berlin, Germany, 11–14 September.

Schinnenburg, M., Debus, F., Otyakmaz, A. and Berlemann, L. (2005). A Framework for Reconfigurable Functions of a Multi-Mode Protocol. In *Software Defined Radio Technical Conference*, SDR'05, Orange County CA, USA, 14–18 November.

SDRFORUM (2005). Available from: http://www.sdrforum.org (June 2005).

Shubrik, M. (1982). *Game Theory in the Social Sciences*. Cambridge, MA, USA and London, UK: The MIT Press.

Siebert, M. and Walke, B. (2001). Design of Generic and Adaptive Protocol Software (DGAPS). In *3Gwireless '01*, San Francisco CA, USA, June.

Smith, M., Welty, C. and McGuinness, D. L. (2003). OWL Web Ontology Language Guide, W3C Candidate Recommendation [online]. World Wide Web Consortium. Available from: http://www.w3.org/TR/2003/CR–owl–guide–20030818/ (June 2005).

SOCRATES (2008). Available from: http://www.fp7–socrates.org/ (September 2008).

SRI (2008). Available from: http://xg.csl.sri.com/index.php (September 2008).

Srinivasan, V., Nuggehalli, P., Chiasserini, C. F. and Rao, R. R. (2003). Cooperation in Wireless *Ad Hoc* Networks. In *22nd Conference of the IEEE Communications Society*, INFOCOM 2003, San Francisco, USA, 30 March – 3 April.

Tuttlebee, W. (2002a). Software Defined Radio: Enabling Technologies. Chichester, UK: John Wiley & Sons, Ltd.

Tuttlebee, W. (2002b). *Software Defined Radios: Origins, Drivers and International Perspectives*. Chichester, UK: John Wiley & Sons, Ltd.

Valletti, T. M. (2001). Spectrum Trading. Telecommunications Policy. Available from: http://www.ms.ic.ac.uk/tommaso/spectrumtrading.pdf.

Von Neumann, J. and Morgenstern, O. (1953). *Theory of Games and Economic Behavior*. Third Edition. Princeton, NJ, USA: Princeton University Press.

Walke, B. (1978). *Realzeitrechner–Modelle*. Munich, Germany: R. Oldenbourg Verlag.

Walke, B. (2002). *Mobile Radio Networks*. Second Edition. New York, USA: John Wiley & Sons, Inc.

Walke, B. H., Mangold, S. and Berlemann, L. (2006). *IEEE 802 Wireless Systems: Protocols, Multi–hop Mesh/Relaying, Performance and Spectrum Coexistence*. New York, USA: John Wiley & Sons, Inc.

Weinberger, D., Gill, J., Hendricks, D. and Reed, D. P. (2005). An Open Spectrum FAQ [online]. Available from: http://www.greaterdemocracy.org/OpenSpectrumFAQ.html (July 2005).

Weiss, T. A. and Jondral, F. K. (2004). Spectrum Pooling: An Innovative Strategy for the Enhancement of Spectrum Efficiency. *IEEE Communications Magazine*, **42** (3), S8–S14.

Werbach, K. (2002). Open Spectrum: The New Wireless Paradigm [online]. Available from: http://werbach.com/docs/new_wireless_paradigm.htm (July 2005).

WI-FI-ALLIANCE (2008). Available from: http://www.wi-fi.org (December 2008).

Wilkins, D., Denker, G., Stehr, M.-O., Elenius, D., Senanayake, R. and Talcott, C. (2007). Policy-based Cognitive Radios. *IEEE Wireless Communications*, **14** (4), 41–46.

WIP (2008). Available from: http://www.ist–wip.org/ (September 2008).

Yaiche, H., Mazumdar, R. R. and Rosenberg, C. (2000). A Game Theoretic Framework for Bandwidth Allocation and Pricing in Broadband Networks. *IEEE/ACM Transactions on Networking*, **8** (5), 667–677.

Yu, W. (2002). Competition and Cooperation in a Multiuser Communication Environments. Thesis (PhD), Department of Electrical Engineering, Stanford University, CA, USA.

Yu, W., Rhee, W., Boyd, S. and Cioffi, J. M. (2004). Iterative Waterfilling for Gaussian Vector Multiple-access Channels. *IEEE Transactions on Information Theory*, **50** (1), 145–152.

Zhao, Q., Tong, L. and Swami, A. (2005). Decentralized Cognitive MAC for Dynamic Spectrum Access In *1st IEEE International Symposium on New Frontiers in Dynamic Spectrum Access Networks*, DySPAN2005, Baltimore MD, USA, 8–11 November.

Zhong, Z., Mangold, S. and Soomro, A. (2003). Proposed Text for Medium Sensing Measurement Requests and Reports. *IEEE Working Document* 802.11–03/340r1.

Index

2.4 GHz   8, 10, 17, 21, 26, 30, 52, 55, 71
3.6 GHz   10, 60–2
5 GHz   9, 13–17, 26, 29, 55, 61, 78–87, 161, 192

Access Category (AC)   195–7, 201–3
Access point (AP)   34, 37, 44, 48, 52, 54–8, 68, 73,
   79, 146, 159
Acknowledgement (ACK)   55, 58, 60, 75–7, 80, 146,
   195, 202–3
Ad Hoc On Demand Distance Vector (AODV)   59
ALOHA   81, 89
Arbitration Interframe Space (AIFS)   50, 195–8,
   201–2
Arbitration Interframe Space Number (AIFSN)
   195–7, 201
Association of Radio Industries and Businesses
   (ARIB)   13

Bandwidth (BW)   13, 17, 21, 28, 37, 46, 52, 54, 62–5,
   66–7, 73, 80–2, 151, 191
Base Station (BS)   16, 32, 44, 61–6, 78–82, 146,
   150, 159, 187
Base Station Hybrid Coordinator (BSHC)   32, 78
Basic Service Set (BSS)   52, 56, 72–3
Beacon   35, 48, 66–7, 92, 94, 148–9, 174
   broadcast   37, 52, 159, 171
   denial   148–54
   dual   149, 155–7
   frequency   156
   grant   37, 148–54
   interval   80, 92, 159
   period(ic)   162, 174–6
   protocol   66
   report   48
Beaconing   5, 38, 67, 145–9, 151, 154–7
Beamforming   51
Behavior   42, 47, 61, 83, 86, 88, 91–3, 101–43, 155,
   173, 191–2, 203–4, 345

Beyond Third Generation (B3G)   1, 43
Block Acknowledgement Request (BAR)   53
Bluetooth   2, 8, 18, 31, 33, 52, 71–2
Broadband Wireless Access (BWA)   62
Broadcast Channel (BCH)   21, 35, 160, 187
Bundesnetzagentur (BNetzA)   11–12
Burst Control Header (BCH)   64, 66
Busy tone   35, 111–14

Capacity   3, 34, 43–4, 81–2, 94, 97, 101, 108–9, 113,
   130, 133, 146, 153, 166–7, 172, 181–2, 210
Carrier Sense (CS)   52, 58, 199, 202
Carrier Sense Multiple Access (CSMA)   34, 52, 74–5,
   110, 162
Cellular   1, 20, 23, 36, 43, 68, 146–7, 205–7, 210
Central Controller Hybrid Coordinator (CCHC)   32
Channel reservation   54, 60, 74, 77
Channel State Information (CSI)   51
Clear Channel Assessment (CCA)   48–50, 58, 74
Clear-to-Send (CTS)   54, 60, 75–80, 195, 199,
   200, 202
Clear-to-Switch (CTX)   60
Code Division Multiple Access (CDMA)   33
Coexistence   1, 4, 14–18, 23, 27, 30–2, 39, 45–7, 56,
   58, 62–72, 78–82, 87–90, 157, 172–80
Coexistence Beacon Protocol (CBP)   66
Cognition circle   189–90
Cognitive Pilot Channel (CPC)   43, 47
Cognitive Radio Architecture (CRA)   43
Collision   52, 54, 75–7, 93, 108, 112–13, 160–3,
   172–6, 180, 201–2
Collision Avoidance (CA)   34, 52, 75–6, 110, 162
Common Channel Framework (CCF)   60
Common Spectrum Coordination Channel (CSCC)
   31, 33, 160
Complementary Cumulative Distribution Function
   (CDF)   90, 130–5, 178–82
Conference Preparatory Group (CPG)   11